NO LONGER PROPERTY
SEATT
D0768306

THE PACIFIC NORTHWEST
Native Plant Primer

THE PACIFIC NORTHWEST

Native Plant Primer

225 Plants for an Earth-Friendly Garden

**KRISTIN CURRIN AND
ANDREW MERRITT**

Timber Press
Portland, Oregon

Frontispiece: Balsamroot (*Balsamorhiza* sp.) and lupine (*Lupinus* sp.) covering a hillside with spring blooms.

Page 5: Glacier fleabane (*Erigeron glacialis*) attracts pollinators of all shapes and sizes.

Page 6 (clockwise from top left): salmonberry (*Rubus spectabilis*), Engelmann spruce (*Picea engelmannii*), sticky purple geranium (*Geranium viscosissimum*), western maidenhair fern (*Adiantum aleuticum*), Columbia windflower (*Anemone deltoidea*), squirreltail (*Elymus elymoides*)

The information in this book is true and complete to the best of our knowledge. All recommendations are made without guarantee on the part of the authors or Timber Press. The authors and publisher disclaim any liability in connection with the use of this information. In particular, eating wild plants is inherently risky. Plants can be easily mistaken and individuals vary in their physiological reactions to plants that are touched or consumed.

Copyright © 2023 by Kristin Currin and Andrew Merritt. All rights reserved.

Published in 2023 by Timber Press, Inc., a subsidiary of Workman Publishing Co., Inc., a subsidiary of Hachette Book Group, Inc.
1290 Avenue of the Americas
New York, NY 10104
timberpress.com

Printed in China on paper from responsible sources
Text design by Laura Shaw based on a series design by Debbie Berne
Cover design by Vincent James and Adrianna Sutton based on a series design by Amy Sly

Library of Congress Cataloging-in-Publication Data

Names: Currin, Kristin, author. | Merritt, Andrew, author.
Title: Pacific Northwest native plant primer : 225 plants for an earth-friendly garden / Kristin Currin and Andrew Merritt.
Other titles: 225 plants for an earth-friendly garden
Description: Portland, OR : Timber Press, 2023. | Includes index. | Summary: "The book will feature 225 northwestern natives that are the easiest for the home gardener to find and grow. Introductory chapters will address the why and how of growing native plants. Two-thirds of the book will consist of the plant entries that focus on the must-have information that readers are looking for. The geographic area covered will be Oregon, Washington, and southern British Columbia, on both sides of the Cascades"—Provided by publisher.
Identifiers: LCCN 2022008024 | ISBN 9781643260716 (paperback)
Subjects: LCSH: Endemic plants—Northwest, Pacific. | Native plant gardens—Northwest, Pacific | Native plant gardening—Northwest, Pacific
Classification: LCC SB439.24.N673 C87 2023 | DDC 635.9/51795—dc23/eng/20220225
LC record available at https://lccn.loc.gov/2022008024

To a resilient future founded on diversity.

Contents

Introduction

Within each of us lies a child. A child wanting to play outside with friends turning over rocks and leaves to find roly-polies and worms, all the while looking for tree frogs, bird's nests, and butterflies. A child who wants to spend the day by a colorful field of wildflowers watching curious creatures fly by under puffy clouds as grasses bend gracefully in the breeze. An inner child longing to wander the forest finding wild, edible fruits that melt in our mouths or play by the river's edge watching fish dart about the watery shallows while making boats from autumn-colored leaves to float out into the stream.

If you grew up on this planet, somehow, in some way, whether you realize it or not, you have a connection to native plants. Maybe it is the memory of an oak tree you climbed as a child or the pungent smell of desert parsley in the warming spring air that strikes a chord with you, whatever the experience, the connection is deep and visceral. Although many of the reasons to garden with native plants can be very adult, the experience can be childlike, bringing us back to a simpler moment in life when we had time to observe and explore, expanding our understanding of the world around us. Set aside for a moment the myriad alarm bells ringing throughout ecosystems for the need to change our predominant gardening practices and remember the simple joy of connecting to nature. Let us keep in mind not only the reasons why we should, but the reasons why we want to connect to the world around us and give ourselves and our gardens the space to do just that. The more you cultivate and nurture a thriving ecosystem in your backyard using native plants as the backbone of your endeavor, the more your childlike curiosity grows as you learn about the species your landscape supports.

◀ Oregon sunshine (*Eriophyllum lanatum*) supports pollinators like small butterflies.

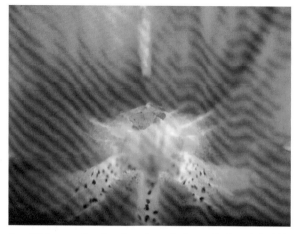

A youthful perspective of great purple monkeyflower (*Erythranthe lewisii*). Photograph taken by the authors' niece, Hazel Womack, age 8.

Native plants shape the landscape as much as they are shaped by it.

We are fortunate to live in the Pacific Northwest where the grandeur and diversity of the natural landscape provides us with endless opportunities for awe and inspiration. We live among temperate rainforests and sun-baked sagelands, alpine meadows, verdant valleys, stately oak savannas, and expansive grasslands. We hope this book helps you cultivate the diversity of the region in your backyard, front yard, community space, pots on your porch, parking strip, or around the farm and inspires your youthful interest in the natural world around you. While giving you reasons to garden ecologically and the basic tools and knowledge to implement your goals, we also want to inspire you to see the beauty and intrigue of Pacific Northwest native plants. There is an amazing diversity in the region and now is the time to cultivate it and enjoy it both as adults and the child within.

Native Plants: What Are They and Why Garden with Them

Native plants give us a sense of place and a heartfelt connection to nature. They characterize the natural landscape, mark the seasons, and attract wildlife iconic to an area. Native plants are defined as occurring naturally in the geographical region in which they evolved, thus shaping the landscape as much as they are shaped by it. They have coevolved with local insects, birds, mammals, and soil biota and have the ability to support the diversity of life that gives the natural world its beauty and its function.

Healthy, functional habitats are becoming increasingly fragmented, lost to development and growing pressure from introduced invasive species. Native plant populations are in turn threatened, as are the species that depend upon them. Not only do insects and wildlife need native plants to thrive but, directly and indirectly, so do we.

It is time for our gardens to complement rather than compete with nature. Gardeners are increasingly utilizing

With native plants gardeners can cultivate a landscape that adds to the local ecology.

the value of native plants in creating beautiful landscapes that support biodiversity, conserve water, and thrive without fertilizers and pesticides. By minimizing lawns and the use of exotic ornamentals and maximizing the cultivation of diverse native plant landscapes, gardeners are able to have a meaningful impact on habitat and species loss while enjoying an attractive, more economical landscape.

Native plants can be used to create perennial beds riotous with seasonal color and pollinator gardens visited by bees, butterflies, and hummingbirds. If you want to grow a food forest out your back door or turn a dry, sunny parking strip into a xeric oasis, plant native plants. For creating rain gardens to filter storm water or establishing stable canopies that hold and build soils with roots that communicate with and benefit soil systems, native plants are key. To cultivate a landscape that adds to the local ecology rather than diminish it, native plants are the way to grow the future.

Pacific Northwest Habitats

The Pacific Northwest is highly varied in geography, geology, and climate, which creates an incredible amount of habitat diversity and variation in the region's flora. Those who live here will never be botanically bored as there are thousands of plants adapted to live and thrive in the multitude of habitats found in the region. Characterized by wet coastal forests along its dramatic coastline, as well as arid high deserts in its interior, the Pacific Northwest has multiple mountain ranges, snow-capped volcanic peaks, island archipelagos, deep canyons, and gorges that add to the iconic imagery of the area.

Geographic features shape the climate in the Pacific Northwest and rainfall plays a role in demarcating habitats. Annual precipitation rates range from less than 10 inches to well over 150 inches a year. In some areas, differences in climate can vary greatly over a relatively short distance. Wet ocean air forced against mountains rises, cools, and releases precipitation, creating a dry region on the leeward side called a rain shadow. This weather pattern, termed orographic lift, makes areas west of the Cascade Mountains generally wetter than eastern interior ones. The Pacific Northwest has the reputation of being a wet place, but the arid interior and dry summers region-wide caused by the interchange of high- and low-pressure systems over the Pacific Ocean challenges that notion.

◀ A multitude of habitats in the Pacific Northwest provides for an immensely diverse flora.

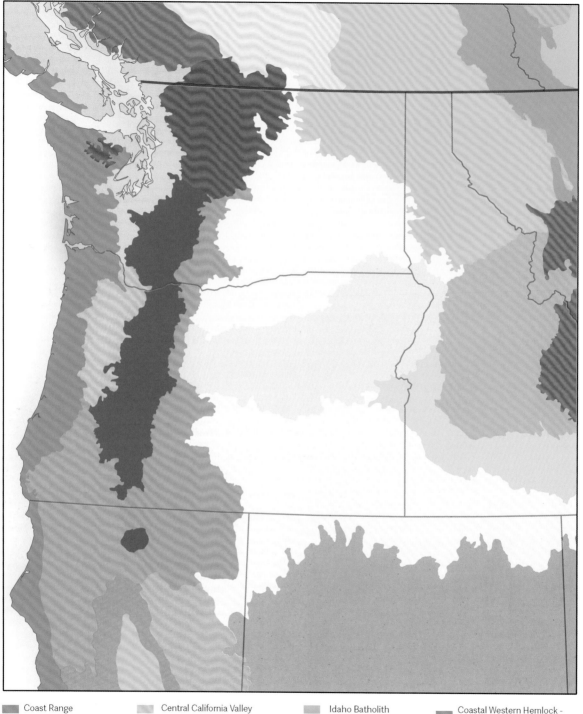

Coast Range	Central California Valley	Idaho Batholith	Coastal Western Hemlock - Sitka Spruce
Puget Sound Lowland	Eastern Cascades, Slopes, Foothills	Middle Rockies	
Willamette Valley	Columbia Plateau	Canadian Rockies	Pacific Ranges
Cascades	Blue Mts.	NW Glaciated Plain	Thompson - Okanogan Plateau
Sierra Nevada	Snake River Plain	Northern Glaciated Plains	
Central California Foothills and Coastal Mts.	Central Basin and Range	North Cascades	
	Northern Rockies	Klamath Mts. / California High North Coastal Range	
		Northern Basin and Range	

There are a number of ecoregions that occupy the Pacific Northwest. Ecoregions are areas where the type, quality, and abundance of environmental resources are much the same. Resources such as water and soil are relatively similar throughout an ecoregion as are climatic conditions and the ecosystems found there. The U.S. Environmental Protection Agency and the Commission for Environmental Cooperation use four levels of increasing complexity to classify and delineate ecoregions in North America. At the first level there are 12 ecoregions in the continental United States, which expands to 967 at the fourth level. The map (opposite) uses the third level of ranking.

Habitats can be variable across ecoregions and intergrade to create a complex web. Understanding your ecoregion, local habitats, and the microhabitats you create in your backyard is the first step in proper plant selection. There are many ways to classify habitats in the Pacific Northwest, but for the purposes of this book we use a simplified classification that we feel translates most effectively for gardeners.

Coastal

A broad term for a network of distinct, smaller habitats that occur on land close to the ocean. This spans sandy beaches and rocky headlands to tidal marshes and coastal forests. These are generally some of the wettest areas of the region, except in parts of the Puget Sound within the Olympic rain shadow. Plants that grow in coastal areas may be adapted to sandy soils, saline conditions, and high winds. Forests tend to be dense and evergreen shrubs and ferns grow thickly. Mild winter temperatures and water availability in both the soil and air encourages rapid plant growth and establishment.

Plants that grow in coastal areas may be adapted to sandy soils, saline conditions, and high winds.

◀ Level III ecoregions of the Pacific Northwest designated by the U.S. Environmental Protection Agency and the Commission for Environmental Cooperation.

Trees can reach sizable proportions in westside forests.

Lowland mixed hardwood-conifer forests are home to the lush growth of moisture-loving species.

Westside Forest

A general term for forested areas west of the crest of the Cascade Mountains that receive ample amounts of seasonal rainfall. These forests tend to have dense understories and canopy cover typically dominated by Douglas fir, western hemlock, Sitka spruce, and western red cedar. Drier, low-elevation areas may be dominated by Douglas fir, Oregon white oak, and Pacific madrone. Water availability and relatively mild winter temperatures promote growth and trees can reach sizable proportions in these forests.

Lowland Mixed Hardwood-Conifer Forest

Widespread on valley slopes and terraces in low-elevation areas west of the Cascade Mountains. Characterized by a codominant mix of conifers and hardwoods such as big-leaf maple and red alder. An assortment of understory shrubs (many of them fruit bearing), ferns, and forbs are found here. This habitat receives ample amounts of seasonal rainfall; however, summers are typically dry as they are across the entire region.

Pine trees are one of the dominant species in eastside forests.

Oak woodlands are one of the region's most threatened habitats.

Eastside Forest

Refers generally to forests east of the crest of the Cascade Mountains that, in comparison to their westside counterparts, are drier, receiving less annual rainfall. Canopy cover tends to be more open, as do understories. These forests are dominated by a mix of conifers such as ponderosa pine, lodgepole pine, and Douglas fir. Grasses, flowering shrubs, and ephemeral wildflowers are denizens of this habitat type. Summers are hot and dry. These forests are adapted to and depend on fire.

Oak Woodland/Oak Savanna

Some of the Pacific Northwest's most threatened habitats are the Oregon white oak (*Quercus garryana* var. *garryana*) woodlands and savannas. Oak savannas are grasslands with a few oak trees per acre scattered among them. An oak woodland has denser tree cover and may intergrade with other forest types and trees such as ponderosa pine and Douglas fir. Oak woodlands tend to be found on drier sites and have a fairly open canopy and understory. These are havens for wildflowers and numerous species of wildlife that rely on oak trees for food and shelter. Oregon white oak is slow growing and adapted to fire, which helps limit encroachment from faster-growing conifers.

▲ Meadows are true havens for wildflowers. ▶ Alpine and subalpine habitats have the coldest climates in the region.

Meadow/Grassland/Prairie

Sunny, open areas with minimal woody growth present. These are biologically fertile habitats comprised of a diversity of perennial bunchgrasses and flowering plants. They may have ample moisture in winter and spring when it rains but may be very dry in summer. Although the plant composition of meadows can differ depending on rainfall, elevation, proximity to a water source, and ecoregion, plants that live in this habitat type are generally adapted to dry summers and ample sunlight along with periodic disturbance such as fire, which limits woody growth. Many of these areas have been developed or converted to agricultural uses.

Drought-tolerant wildflowers in shrub-steppe habitat take advantage of seasonal moisture by blooming bountifully in spring.

Shrub-steppe

Arid areas comprised of drought-tolerant shrubs such as sagebrush, bitterbrush, and rabbitbrush, as well as perennial bunchgrasses, wildflowers, and cryptobiotic soil crusts of lichens and mosses. Fire and drought play an important role in this habitat type. Plants that grow here are adapted to seasonal drought and extreme temperatures. Covering a significant portion of land in southeastern Washington, southeast Oregon, and southern Idaho, large areas of intact shrub-steppe habitat have been lost to development and conversion to other uses such as agriculture and grazing.

Subalpine/Alpine

High-elevation areas with the coldest climates in the region, characterized by a short growing season and extreme conditions. Subalpine areas consist of the forests, meadows, marshes, and rocky outcroppings on high upland slopes below timberline. Alpine habitats occur in the areas above timberline where low-growing plants cling tenaciously to rocky exposures. Water is seasonally available either as precipitation or melting snowpack. Subalpine forests are dominated by a mix of conifers such as Engelmann spruce, mountain hemlock, subalpine fir, and western larch.

Aquatic plants like yellow pond lily (*Nuphar polysepala*) are adapted to being submerged in or floating on top of water.

Wetlands are saturated, sensitive areas vital to the health of many species.

Aquatic

Freshwater ponds, lakes, and streams. For the purposes of this book, this does not include marine or saltwater habitats. Plants that live here are adapted to being submerged in or floating on top of water.

Bog/Marsh/Wetland

Saturated, sensitive areas where water is available year-round. Bogs are places where rainwater collects in depressions and the stagnant saturation results in acidic conditions to which certain plants have adapted. Marshes are flooded areas that remain waterlogged. Wetlands in general are productive places vital to the health of a multitude of species. These areas are disappearing and deserve protection.

Riparian plants are important for holding soils, filtering water runoff, and shading waterways.

Vernal pools are isolated depressional wetlands that are wet in spring but dry in summer.

Riparian

The area of land along waterways and bodies of freshwater, the boundary of which shifts from site to site. Typically, a lush and productive area with a high water table. Soils remain saturated close to the water, but on raised banks and areas farther away from water sources, and if water levels lower seasonally, soils become drier. Plant growth here is important for holding soils, filtering water runoff, and shading waterways.

Vernal Pool

Small, isolated depressional wetlands that are saturated in spring but dry in summer. Plants that grow here are adapted to seasonal flooding and drought. A unique habitat that is threatened in the Pacific Northwest.

Dry, rocky soils are home to drought-tolerant bunchgrasses and perennials.

Pioneer species like fireweed (*Chamaenerion angustifolium*) thrive in disturbed areas such as forests recovering from fire.

Rocky

Characterized by the predominance of rock or rocky soils, these areas have sharp drainage and little water retention. Some showy species of wildflowers are able to thrive in this seemingly inhospitable habitat. In some cases, this is more of a microhabitat found within other habitats, such as a sunny, rocky outcropping in a wet lowland forest.

Disturbed

This could be considered the fastest growing habitat in the Pacific Northwest, with intact habitat increasingly lost to development and conversion to other uses. There are many species of plants, both native and nonnative, adapted to colonize and take advantage of areas of disturbance. These plants are called pioneer species and they may grow and spread quickly.

◀ Some of the region's showiest species of wildflowers thrive in rocky habitats.

Beauty in Biodiversity

Supporting Life in the Garden

The beauty of a garden is much more than just the color and structure of its flowers and foliage. Collectively, we have developed a fixation on form over function in our concept of what is beautiful in our landscaping that limits our perspective on all it can provide. As we come to understand that the rapid increase of human-dominated landscapes has displaced functional habitat resulting in species loss and extinction, it becomes imperative that we begin to see our backyards as part of the local ecology and seek to harbor life other than our own. In doing so, we expand our concept of what is beautiful in the garden to include the artistry and vitality of the vast array of life waiting and needing to share space with us.

Adopting a more ecological aesthetic is rewarding. The pleasure of knowing that our backyards are a sanctuary for biodiversity transforms function into a beautiful quality and a source of pride. A leaf nibbled on by a swallowtail caterpillar becomes a cherished sight rather than an unwanted one. The brilliance of the bloom is as delightful as watching pollinators sip from it. Seeing, hearing, and forming a more direct connection with a diversity of life thriving in our backyards is beautiful and enjoyable.

◀ Buckwheats (*Eriogonum* spp.) attract a beautifully diverse array of life.

Swallowtail butterfly larva on *Lomatium* sp.

Manzanitas (*Arctostaphylos* spp.) support important pollinators such as butterflies and leafcutter bees.

By planting native plants and creating wildlife habitat gardeners can have a positive impact on reducing species loss. Having evolved in tandem with native insects, birds, animals, and soil biota, native plants have developed special supportive and synchronistic relationships with local species that are not fully replicated by exotic plants. While there are nonnative plants that can perform multiple functions in our gardens and may even benefit local life to some degree, most exotic species of plants sever local food webs. Entomology and wildlife ecology professor Douglas Tallamy's research into the impact of nonnative plants on breeding birds reveals that in the Washington, DC, area a native oak tree can support over five hundred species of caterpillars whereas a gingko from Asia supports close to none. This is because most herbivorous insects, including the larvae of butterflies and moths, are specialized to eat only plants with which they coevolved. Considering that a brood of Carolina chickadees requires over five thousand caterpillars during their rearing, we can begin to conceive of the breakdown of local food webs that results from replacing native plants with exotic ones. Planting and maintaining a diversity of native plants in the landscape is vital to providing a haven for local birds, animals, and insects.

Beauty is in the eye of the beholder, and what looks like a lovely landscape to us may be an unattractive one in a bee's eyes. It is time for us to embrace a more holistic perception of what is beautiful in our gardens and start to see the beauty in biodiversity. Turning our gardens into places of refugia is imperative and rewarding, and creating a space where people and nature coexist provides us with an intimate outside classroom where we can learn to appreciate other species and be better stewards of the land we share with them.

Creating Wildlife Habitat

There are four basic elements essential for creating wildlife habitat: food, water, cover, and space to raise young. Providing a sanctuary for wildlife in your backyard starts by planning for these elements and specifically tailoring them to the needs of the local fauna you would like to encourage.

Studies show that to properly provide a place where wildlife thrives, at least 70 percent of the plant species in your garden need to be locally appropriate native plants. Observing and learning what wildlife is local to your area and what plants they utilize and forage upon will help you make proper plant selections. If you plant it, they will come. This phrase may be true so long as you plant appropriately for your ecoregion and the needs of local wildlife. For example, planting host plants for monarch butterfly larvae, milkweed (*Asclepias* spp.), in western Washington will not ensure monarch butterflies visit your garden. Milkweed does not naturally occur west of the Cascade Mountains in Washington, and consequently monarchs are a rare sight in that part of the state. To complement the general information given in this book, there are resources both online and in print that can provide gardeners with more specific regional lists of plants and the species they support. See References and Resources for recommendations and consult local backyard habitat programs. Again, take time to observe what is happening in your garden, learn what plants are native to your area presently and historically, and what local wildlife is around and wanting.

Plant diversely when planning for diversity. Plant an array of native plants that flower and fruit at various times of the year, thus providing continual seasonal forage for wildlife. Planting native vegetation that differs in height and structure provides various types of shelter from which wildlife can choose. Keep in mind that not only trees and shrubs provide cover for wildlife, but so do plants like bunchgrasses and ferns. Planting a diversity of native plants provides a variety of options for resting, nesting, feeding, and breeding.

Creating a sanctuary for wildlife in your backyard means providing food and water, as well as resting and nesting sites.

Showy milkweed (*Asclepias speciosa*) is a host plant for monarch butterfly larvae.

Planting a diversity of native plants provides wildlife with a variety of options for resting, nesting, feeding, and breeding.

▲ Providing a water source for wildlife is important. ◄ To properly provide a place where wildlife thrives, at least 70 percent of the plant species in your garden should be locally appropriate native plants.

Whether it be a nearby natural source, a bird bath, or just a shallow bowl kept full, providing water is essential along with a source of food and shelter.

There are cultivation and maintenance techniques that benefit wildlife but may deviate from standard garden and landscape practices. These include leaving leaf litter, not removing dead trees if they pose no safety hazard, creating brush piles if your space is large enough to accommodate them, leaving nurse logs, and allowing last season's growth to remain in the garden over the winter. Insects and animals utilize these materials for nesting and winter protection. Trimming of plants should be timed so as not to remove needed forage and nesting materials or disturb nesting and overwintering wildlife. Find the balance between form and function as you maintain the garden for both your needs and those of the creatures that visit. You can keep it pretty, but more importantly, keep it productive.

Gardeners preparing and maintaining backyard habitats must understand it is critical they not use pesticides, herbicides, or other chemical products in the garden as these can hurt or kill wildlife. It is also important to never buy plant stock that has been treated with neonicotinoids, a widely used class of insecticides available commercially under many different names, or any other systemic insecticide. Systemic insecticides are absorbed by the plant and can be present in pollen and nectar, making them particularly harmful to bees and other pollinators. Utilizing organic gardening methods is essential when creating healthy habitat for wildlife. Cultivate a chemical-free garden that can safely harbor and attract beneficial insects and other pest predators, thereby generating a natural form of pest control in the garden.

Once you have planted your wildlife-attracting garden, you may need to protect plants while they become established. Take steps such as caging young trees and using floating screen covers for young plants. Protect seeded beds from foraging birds and small animals with covers such as Reemay and chicken wire. Safely deter wildlife and pets from damaging the garden as it gets rooted so everyone can reap the rewards when plants mature.

For the Birds

Birds fill the air with song and color and are a seemingly ubiquitous joy, yet they need our help. Bird populations in North America are in decline across all habitat types due to human-related activities such as development, agriculture, and pesticide use. Studies conducted by the Audubon Society reveal two-thirds of North American bird species are facing the risk of extinction due to climate change. Birds act as pollinators, seed dispersers, and eat insects we might consider garden pests; they are key players in a functional ecosystem and are indicators of its health. It is time to turn our backyards into places where birds can not only land and find shelter but where they can also find the food they need to survive and thrive.

Birds are key players in a functional ecosystem but bird populations in North America are in decline across all habitat types due to human-related activities.

The health of bird populations is linked to the health of native plant populations and native insects such as butterflies and moths.

Golden currant (*Ribes aureum*) attracts cedar waxwings with its sweet, edible fruits.

Native plants provide birds with food mainly in the form of fruits, seeds, and insects. Most of the birds we encounter in our gardens eat insects and need them to rear their young. Even birds we think of as seed or nectar feeders, such as hummingbirds, eat insects and feed them to their chicks. Native plants are needed to support native insect populations, particularly herbivorous insects such as the larvae of butterflies and moths, because most insect herbivores are specialized to eat only plants with which they share an evolutionary history. Therefore, the health of bird populations is linked to the health of both native insect and native plant populations.

Caterpillars are essential food for nesting birds and studies by Douglas Tallamy and his colleagues show native woody plants support thirty-five times more caterpillar biomass than nonnative woody plants. Planting key species that host large numbers of Lepidoptera (butterflies and moths) such as oaks (*Quercus* spp.), wild cherries (*Prunus* spp.), willows (*Salix* spp.), and species of *Ceanothus* is a great way to support butterflies and feed birds and their chicks at the same time. Look forward to seeing a leaf nibbled here and there by caterpillars but know that a healthy bird population provides a constant predatory patrol of insects that might otherwise be considered garden "pests."

The Pacific Northwest is home to many trees, shrubs, and perennials that bear fruits birds relish in summer and fall. Incorporating fruit-bearing plants such as currants, wild cherries, serviceberries, elderberries, and others into the garden provides food for birds and can be rewarding for the gardener as both the fruits and flowers of these plants can be showy and attractive. Fruits may also be edible to humans, and the more delicious the fruit the harder it may be to share but leave some for the birds!

Planting a variety of flowering native perennials and annuals and letting them go to seed provides a natural food source for graniverous (seed-eating) birds. Graniverous birds may be particularly attracted to the seeds of sunflowers (*Helianthus* spp.) and their relatives such as balsamroots (*Balsamorhiza* spp.), as well as other members of the family Asteraceae. Be aware that sowing wildflower seed can be the same as putting out birdseed, take steps to protect seeded beds from foraging birds until seedlings are established.

Birds need places to take cover and build nests to raise their young. Planting a variety of plants with various heights and structures provides birds with options for shelter. Birds also use many parts of plants for nesting material. Leaving fibrous matter produced by forbs (herbaceous

Oregon white oak (*Quercus garryana* var. *garryana*) provides many species of birds, such as this pair of pileated woodpeckers, with shelter and a place to raise their young. Oaks also support populations of butterflies and moths, which make up an essential part of birds' diets.

Gardeners can greatly benefit bird populations with their plant choices. Planting native plants is essential as they support the food webs that sustain local bird populations. By planting native plants, gardeners can create bountiful bird habitat in their backyards and become part of a collective corridor that feeds and shelters our feathered friends.

Supporting Native Pollinators

The decline of pollinators is a scary universal reality, but gardeners should know they are uniquely positioned at the forefront of being able to turn fear into hope. Many hundreds of species of native bees in North America and Hawaii are in serious decline, threatened with extinction, or have vanished altogether. The situation is a silent but serious crisis with multiple species of bees now listed under the U.S. Endangered Species Act. There are myriad kinds of pollinators including bees, butterflies, moths, bats, birds, and even flies and mosquitoes. Pollinator populations in general are declining due to loss of habitat, pesticide use, introduced diseases, and climate change. With over 85 percent of the world's flowering plants requiring pollinators

flowering plants) and bunchgrasses in the garden rather than removing it when plants go dormant provides birds with nesting supplies. For example, birds may strip fiber from the stalks of milkweed left uncut over the winter for cordage to bind their nests in the spring, and hummingbirds line their nests with the downy fluff of cattail seeds produced the year before. Keeping the garden too clean and tidy removes materials that benefit birds, as well as the insects they prey upon. And don't forget the bird bath! A source of water is vital for birds and other wildlife including insects.

Native bees collecting pollen and nectar from balsamroot (*Balsamorhiza* sp.) in a xeric garden. By turning your garden into a sanctuary for native pollinators you can help stem the tide of species extinction.

A common yellowthroat (*Geothlypis trichas*) on tall Oregon grape (*Berberis aquifolium*) in a community wetland planting for pollinators and wildlife.

This community wetland planting benefits wildlife, birds, pollinators, and people!

Nonnative invasive weeds once smothered this wetland area in a neighborhood in northern Oregon. Community members devised a plan to remove the noxious weeds and replant the area with locally appropriate native plants to benefit wildlife, birds, pollinators, and the overall function and health of the wetland. Abundant habitat full of native plants now thrives there, enriching and beautifying the neighborhood. A boon to species such as this common yellowthroat, which prefers thick vegetation, particularly in wetland areas, where they build hidden nests on or near the ground. Although this species is abundant overall, its populations have declined by about 38 percent since 1966 and suffer from habitat loss and pesticide use. Some nonmigratory populations of this bird in California and

Texas have declined to the point of nearly becoming locally extinct, also known as extirpation. Impacts to a species as a whole that result from extirpation are reductions in range, genetic diversity, and population size, which increases the likelihood of the complete extinction of the species. An insectivorous bird, common yellowthroats glean insects off plants, thus controlling infestations of herbivorous insects, benefiting not only the community planting but also neighboring gardens as well. A great example of a truly reciprocal, beneficial relationship between gardeners and wildlife. For more information about this and other birds you might find in your garden visit the Cornell Lab of Ornithology's website allaboutbirds.org.

for their reproduction and to produce the crops, fruits, and seeds that feed the world, the situation becomes dire. Gardeners can help stem this tide by turning whatever space they can into a sanctuary suitable to support and sustain native pollinators.

The honey bee is the poster child of pollinators. While the honey bee is facing population declines that are concerning to all, especially given they provide us with a formidable pollinating army for farms and orchards, it is important to keep in mind that honey bees are not native to the Pacific Northwest. Honey bees may even outcompete native pollinators and have an effect on the composition of native plant communities. There are approximately 4,000 species of bees native to the United States and Canada and they come in a range of shapes and sizes. Many native bees lead solitary rather than communal lives and are not aggressive or able to sting. Native bees are better adapted to local climates and can tolerate colder and wetter weather than honey bees, providing critical pollination services to both wild and cultivated plants even in inclement conditions. By focusing efforts on supporting native bees and other native pollinators, gardeners can help stem the tide of local species extinction and support the pollinators best adapted to supporting native plant communities.

There is a diverse list of native pollinators that may visit your garden and their needs may differ. However, there are a few universal elements necessary for providing native pollinator habitat: continual forage, places to rest and nest, and protection from pesticides.

Planting an array of plants that bloom at different times of the year is important and provides pollinators with a steady source of nectar and pollen throughout the growing season. Choose plants with various flower shapes, sizes, and colors to help provide options to fit the various shapes and sizes of native pollinators themselves. Hummingbirds favor long tubular flowers like those produced by orange honeysuckle (*Lonicera ciliosa*), while other pollinators prefer more easily accessible flowers like Oregon sunshine (*Eriophyllum lanatum*). Large bumble bees seek out penstemons that have large flowers, just as small bees

There are myriad kinds of native pollinators.

Bombus nevadensis, one of our largest native bumble bees, on golden currant (*Ribes aureum*).

► Hummingbirds love the long tubular flowers of orange honeysuckle (*Lonicera ciliosa*).

▼ Having a garden full of flowers with many colors, shapes, and sizes will attract the greatest array of pollinators.

swarm to penstemons with small flowers. Red-colored flowers may not attract bees, yet could be a magnet for hummingbirds. Grouping similar species in patches will help pollinators find their favorite blooms. Having a garden full of flowers with many colors, shapes, and sizes will attract the greatest array of visitors.

Most native bees are solitary and nest either in tunnels underground, holes in snags and trees, or the soft pithy stems of plants like elderberries (*Sambucus* spp.), blackcap raspberry (*Rubus leucodermis*), and goldenrod (*Solidago* spp.). Leave patches of bare earth in and near your garden for ground-nesting bees to create burrows for eggs and overwintering queens. Be aware that tilling the ground can disrupt, dislodge, or kill developing or hibernating bees. When trimming plants leave some of the pithy branches and stalks of shrubs and perennials for cavity-nesting bees (8 to 24 inches of flower stalks or canes, 4 to 6 inches above branch or leaf nodes of shrubs or trees). Gardeners can combine function with art by creating "bee hotels" (particularly useful for mason bees), which can be an interesting and attractive feature in the garden and help provide supplemental nesting options. Having a water source is also important and can be as simple as a small dish of water placed somewhere in or near the garden that pollinators can access safely and frequently. Make sure to put stones or sticks in the water as bees need solid footing for landing and takeoff and will drown if water sources do not provide them with a safe approach and exit.

Grow organically and protect pollinators and pollinator habitat from pesticides and other toxic garden chemicals. Particularly perilous for pollinator health are systemic insecticides, such as neonicotinoids, which can be absorbed by plants and become present in pollen and nectar. Be careful when sourcing plants and do not buy plant stock treated with insecticides.

Another important thing to keep in mind when sourcing native plants for the pollinator and wildlife garden is the difference between native plant cultivars and true native plants. Cultivars are plants that have been bred or

The pithy stems of plants like blue elderberry (*Sambucus cerulea*) provide habitat for cavity-nesting bees such as mason bees.

Grow organically and protect pollinators from pesticides.

Cultivars like this red-leaved ninebark may be pretty but not beneficial, and even potentially harmful, to native insects.

Turn your garden into a pollinator paradise.

selected for qualities such as flower size and shape, or leaf color. Examples of this are double-flowered mock orange, such as *Philadelphus* 'Buckley's Quill', and red-leaved ninebark, such as *Physocarpus opulifolius* 'Monlo'. Cultivars may be pretty but may not benefit native pollinators and insects. They may be sterile, have reduced nutritional value, increased toxicity, or flower structures too complex for pollinators to navigate easily. Native plant cultivars may be the only options carried by some nurseries and may even be advertised as beneficial to pollinators. Whether native plant cultivars benefit native pollinators and wildlife depends on the individual plant and its attributes. Gardeners should remain savvy to this, choose plants wisely, and always favor planting true native plants.

Supporting native pollinators goes hand in hand with supporting native insects and in turn the animals that prey upon them. More butterflies and moths equals more birds! Many of the basic principles of creating habitat for wildlife applies to creating pollinator habitat. Plant a diversity of locally appropriate native plants, reduce lawns and the use of exotic plants, leave the leaf litter, and don't keep the garden so tidy as to create an environment too sterile for biodiversity to thrive. Offer nesting places, provide a water source, and don't use chemical products in the garden. Supporting native pollinators is a gateway to sustaining natural food webs in general, and creating a pollinator paradise is a powerful and hopeful step that gardeners can take toward ensuring a better future for all.

◀ Pollinators such as swallowtail butterflies love the flowers of mock orange (*Philadelphus lewisii*), but double-flowered cultivars of this plant may produce flowers that are confusing or inaccessible to pollinators.

Plant Selection and Cultivation

Native plant gardens and landscapes can be both beautiful and economical. If appropriately sited and adapted to the ecoregion and soils they are planted in, native plants do not need fertilizers, an expensive input that more conventional gardens may require. Appropriately sited and established native plants may also be able to thrive without irrigation, reducing water use and costs. A healthy backyard habitat will not need or want the use of costly pesticides as it seeks to attract beneficial insects and natural garden pest predators who maintain the garden for free. Collective, connective backyard sanctuaries for birds and pollinators nurture and support the animals and insects that benefit and pollinate our crops, providing us with food and a thriving "eco-nomy." For gardeners ready to reap the rewards of creating a beautiful and bountiful native plant garden or landscape, there are a few basic things to know when selecting, planting, and maintaining native plants.

◀ Appropriately sited and established native plants often thrive without irrigation, reducing water use and costs.

Right Plant, Right Place: Choosing the Right Plant for You and Your Garden

Selecting the right plants for your site is an important step in cultivating a native plant garden that thrives with minimal maintenance and inputs once established. Using plants that occur naturally in and are therefore adapted to your area, as well as understanding your specific site conditions—sunlight exposure, soil type, soil moisture, and climate—and choosing plants that grow in those conditions will lead to happier plants and happier gardeners.

While there are generalist species of native plants that can be found in a diverse range of habitat types and across multiple ecoregions, many plants native to the Pacific Northwest are adapted to living in specific habitats. A species can be comprised of variations or subspecies, which may be distinct in growth habit and physical characteristics, as well as in range and preferred habitat. Selecting locally and ethically sourced plant material and planting the variations of species and local genotypes appropriate for your area is important to both the surrounding ecosystem and the success of your garden.

Variation within a species or genus provides options for gardeners. A plant that grows in wet, shady sites may have a close and similar looking relative that thrives in more arid habitats. For example, Columbia larkspur (*Delphinium trolliifolium*) grows in moist, shady woodlands almost entirely west of the Cascade Mountains, whereas upland larkspur (*Delphinium nuttallianum*) grows abundantly east of the Cascades in drier, sunnier shrub-steppe habitat, rocky meadows, and open pine forests. Gardeners with shady gardens west of the Cascades will struggle and likely fail to get upland larkspur to thrive in their wet woodlands, while gardeners living in hot, sunny, dry areas east of the mountains will have to labor and water regularly to keep Columbia larkspur alive. Switch that around, however, and you have two gardeners in very different settings both with species of *Delphinium* adapted to thrive in their habitat and climate and requiring much less labor, maintenance, and resources. Many of the plant profiles in this book provide suggestions of alternate species for different habitats. Botanical manuals and online plant databases can provide even more information about species, their variations, and ranges.

With irrigation, soil amendments, and the right sunlight exposure, gardeners can create microclimates and microhabitats that differ from those in the surrounding area and may be able to grow plants not found naturally in their locale. When selecting plants for creating backyard habitat and a sanctuary for biodiversity, understand that

Columbia larkspur (*Delphinium trolliifolium*) grows in moist, shady woodlands west of the Cascade Mountains.

Upland larkspur (*Delphinium nuttallianum*) grows in sunny meadows, open pine forests, and shrub-steppe habitat east of the Cascade Mountains.

plants grown outside of their natural range may or may not be able to support local species of native insects and wildlife. Furthermore, plants from a different ecoregion may hybridize with closely related species found locally, such as columbines (*Aquilegia* spp.), which are particularly quick to cross with both cultivated and wild relatives. It is important our gardens become diverse places of refugia while being mindful of their impact on surrounding plant and animal communities. You can protect the genetic integrity of native plants in your area and better support local biodiversity by planting species found naturally where you live.

It can be helpful to select species that typically grow together in the wild. This serves to ensure plants will complement each other, form a better semblance of functional habitat, and require similar watering regimes, making it easier for gardeners to have an irrigation system that works for all the plants in the garden.

As the gardener it is your choice how to blend the shapes and colors from the palette of Pacific Northwest native plants on the canvas of your landscape. Planting species that grow together in the wild does not mean your landscaping has to look wild or unkempt. While creating an environment productive enough to encourage biodiversity, we can still shape the structure and composition of the garden with a layout that accentuates particular qualities of plants. Grouping plants of the same species together, coupling plants with complementary flower colors or differing heights, or siting a silver-leaved specimen in front of a deep

Red columbine (*Aquilegia formosa*) hybridizes easily with both cultivated and wild relatives.

▲ Planting species that typically grow together in the wild helps ensure the plants in your garden will have similar irrigation requirements.
◄ Gardeners can choose how to blend the shapes and colors from the palette of Pacific Northwest native plants on the canvas of their landscape.

green shrub to provide contrast are some of the many ways to bring style and structure to your planting. Pick species that appeal to you, appeal to the wildlife you would like to encourage in your backyard, and are appropriate for your area and the growing conditions in your planting site, then paint a beautiful and bountiful native plant landscape.

Sourcing Native Plants

The first rule of sourcing native plants is do not dig them from the wild! Native plant gardeners should seek to augment native plant communities, not diminish them. Digging plants from the wild is often unsuccessful as many species simply do not transplant well. Furthermore, digging plants from natural areas can harm native plant populations, especially rare, endemic, threatened or endangered ones, and damage both the plants you are digging and plants living nearby. It drains the local ecosystem and the disturbance leaves bare ground that may be colonized by invasive species, hopefully not ones that came in as seed on your clothing or shoes! With growing pressure on native plant communities from development, agriculture, invasive species, and climate change, as well as species suffering population declines from overharvesting, we need to give plants a break and seek more ethical ways to propagate and source native plants than digging them from the wild.

The best way to source native plants and minimize your impact on wild populations is to obtain them from nurseries that maintain ethical standards and practices when propagating and selling native plants. Nurseries should not be digging plants from the wild to sell, the exception to this being salvaging plants to save them from disturbance or development. Seed and cuttings are the most ethical ways to propagate native plants. Ask nurseries about their propagation practices. Are plants seed propagated? Are seed and cutting materials collected ethically and with sensitivity to parent plant population health? Also ask if plants have been sprayed with pesticides and never buy plants treated with systemic insecticides.

Make sure the plants in your garden are ethically sourced and propagated. These young penstemons have been grown from ethically collected seed.

Bachelor's button (*Centaurea cyanus*) is a common component of wildflower seed mixes even though it is not native to the Pacific Northwest. This plant has escaped cultivation and spread aggressively into natural areas, outcompeting and harming native plant communities.

There are nurseries that may sell cultivars of native plants and label them as beneficial to local wildlife and pollinators. See the "Supporting Native Pollinators" section in the previous chapter for more information about cultivars and why they may not be beneficial to native insects and wildlife. Planting true natives propagated from locally sourced seed will provide you with plants best suited to your area and to supporting the native wildlife and insects that live there.

Premade seed mixes are a popular way of planting wildflowers, particularly in meadowscapes, but most commonly sold packets of wildflower seed contain plants that are not native to the Pacific Northwest and may even be noxiously invasive. Bachelor's button (*Centaurea cyanus*) is a prime example of this. Beautiful and able to grow easily in the Pacific Northwest, though native to Europe and western Asia, bachelor's button is a common component of wildflower seed mixes. This nonnative plant has now escaped cultivation and spread aggressively into natural areas where it is outcompeting and harming native plant communities. Studies have shown many commercially available wildflower seed mixes are either partly or entirely comprised of species that are invasive and not native to the region. Do not use wildflower seed mixes such as these. There are a number of seed suppliers both large and small that can provide ethically sourced, regionally appropriate native plant seeds and mixes to gardeners and landscapers.

If there are no nurseries in your area with locally appropriate native plants or if certain species that grow in your area are not available commercially, then gardeners may decide to propagate native plants from seed and cuttings themselves. Follow ethical standards when working with wild populations of native plants. Always properly identify plants and never collect or remove any parts of rare plants. Only collect seed or cuttings from healthy, robust

Planting true natives propagated from locally and ethically sourced seed will provide you with the plants best suited for your area.

these populations could be harmful. Seasonal conditions dictate growth and pollination for all species of plants, be mindful when seed production is below average in a given year and be aware of the impact harvesting seed in those years may have on both plant communities and the animals that depend on the seed as a food source. Collecting seed of annual plants can be particularly impactful to wild populations as annual plants live for only one year and depend on seeds as their ticket to the future. Taking vegetative cuttings for propagation should also be done respectfully as cutting plant material can harm parent plants or spread disease. It is important to properly identify the plants you want to work with, understand their variations, ranges, population size, life cycle, and function in the ecosystem. Have a good understanding of which species are appropriate for wild seed collection, which species are not, and how to collect plant material without harming local plant populations and the species that depend on them.

Propagating by seed can be a meaningful way to learn about the life cycle of native plants. Gardeners may learn it takes more than one year for some seeds to germinate. Seeds of columbines (*Aquilegia* spp.) and Oregon grapes (*Berberis* spp.), for example, may take up to two years to stratify and germinate. Gardeners will also discover certain species, such as some of the most beautiful lilies in the Pacific Northwest, take many years to develop and flower from seed. Learning about the life cycles of plants can lead to a deeper respect and understanding of native plants and how to nurture them both in the garden and in the wild.

Balsamroot (*Balsamorhiza* spp.) presents a good case example. Species of balsamroot are showy and iconic to certain areas of the Pacific Northwest, yet populations face increasing pressure from development and agriculture. It can be exceptionally long lived and have a very deep taproot that grows 5 to 9 feet into the soil. Efforts to dig and transplant it are met with failure and the destruction of the plant. Balsamroot produces viable seed, which is the only ethical way to propagate them for gardens and landscapes. Seed collection needs to be accompanied by proper identification and an understanding of balsamroot species, their abundance, and

populations and have permission from landowners to collect plant material. Gardeners should not need much seed. A small handful of viable seed if stored, sown, and cared for properly can yield many plants. A very important rule of thumb for collecting wild seed is never taking more than 5 percent from the population, although even this fails to suffice as an ethical compass in some cases. Some native plants are too rare to remove plant material from wild populations and those plants may be protected and illegal to collect. In other cases what may be an overall prolific species may have isolated, disjunct populations growing outside of its normal range. These disjunct populations may be small, more sensitive, and potentially on a different evolutionary path as isolation leads to variation. Removing plant material from

Lupine seeds developing.

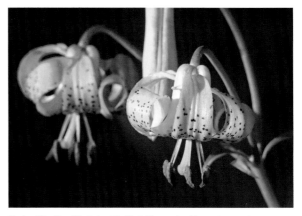

If collecting and sowing your own seed, be sure to follow ethical collection guidelines. Seed collection of annual species like these elkhorns clarkia (*Clarkia pulchella*) can have a big impact on wild parent populations.

Native lilies like this Columbia lily (*Lilium columbianum*) may take many years to develop from seed to flower.

Populations of balsamroot and other native plants face increasing pressure from human development.

Balsamroot (*Balsamorhiza* sp.) can be long lived and develop a deep taproot 5 to 9 feet long.

Do not dig plants from the wild. Populations of species such as these trilliums can be deeply impacted if removed from the wild and must be ethically propagated. Native plant gardeners should seek to augment native plant communities, not diminish them.

limitations. Different species of balsamroot easily hybridize, and it can be difficult to decipher intergrading populations. Additionally, some species of balsamroot are rare either generally or locally. Like sunflowers, the seed is rich in oils, making it an important nutritional food source for birds, insects, and wildlife. When collecting seed take into account the needs of wildlife that depend on the seeds and future plants for food. If there are only a few plants around, it is not ethical to collect any plant material from them. Sown seed will need protection from foraging birds and animals. Cultivation requires patience as balsamroot can take as many as ten years to flower but has huge rewards once plants mature. Watching balsamroot grow, bloom, and attract wildlife and pollinators is rewarding for gardeners who have taken steps to ethically source or propagate the plant. The gardener who has gone through this process will never look at a population of balsamroot the same again and will have gained deep respect for the plant and its role in the ecosystem, as well as its vulnerability.

Sourcing native plants is an ethical journey. Gardeners should make sure they are buying true, locally appropriate, organically raised native plants from nurseries whose propagation practices protect and prioritize the health of parent plant populations. For those gardeners that need or want to propagate their own plant stock, collecting and sowing seed is an option if done properly and respectfully. More specific information about propagating native plants is outside the scope of this book but there is information out there to help gardeners. No matter what, be sure no native plants or animals are harmed in the making of your garden!

Even native plants need a little love and care.

Properly preparing your site for planting can lead to long-term success.

Creating a Native Plant Garden

Not to disappoint the reader, but it is a misconception that native plant gardens are maintenance free. Even native plants respond to and need a little love and care. There are ways, however, to lower the maintenance needs of native plant gardens. Low-maintenance gardens are the result of properly siting plants, preparing the planting site, and establishing newly planted plants. High-maintenance gardens are the result of putting the wrong plant in the wrong place, not properly preparing planting sites, and expecting native plants to thrive immediately without proper establishment. Gardeners who expect all native plants to be invincible are setting themselves up for failure. Help yourself be successful by selecting and sourcing plants suited for your region and following these basic steps when preparing, planting, establishing, and maintaining a native plant garden.

Site Preparation

The more preparation you do to your site, the better your results will be. If you already have a well-developed, healthy garden space to plug natives into or slowly convert to native plants then you are well on your way. If, however, your planting site is full of invasive species and compacted or damaged soils, you have some work to do before planting.

No plant likes compacted soils. Soils need to be broken up and aerated so they are loose and porous enough to allow oxygen into the soil, promote water absorption, and not impede root growth. Try to avoid tilling as this can be harmful to healthy soil biota and ground-nesting beneficial insects. Mixing a little organic matter such as decomposed leaves into the soil can help. Native plants are adapted to local soils and whatever you add to your soil should be in line with what plants grow in naturally. Do not use overly rich compost and avoid manure and anything with fertilizers in it. Most prefab soil mixes in garden stores are too rich for native plants. If using amendments, use them throughout the entire planting area, not just the hole you are planting; doing the latter may restrict root growth. Make sure any amendments used in the garden are organic and free of weed seeds.

Many native plants need good drainage. Gardeners can find instructions online on how to conduct simple tests to determine how well their soils drain. Take steps to provide good drainage to plants that need it. Protecting plants from soggy conditions can involve adding materials that help improve drainage to the soil or planting into raised beds or mounds that help shed excess water.

Introduced invasive plants often produce seeds that remain dormant in the soil for years if not decades. One of the most important parts of site preparation is removing or suppressing weeds and the existing weed seed bank from the planting site. The more thoroughly done before planting, the easier it will be for gardeners to maintain in the future. Depending on how pervasive the weed population is will dictate the removal techniques you employ. Hand pulling is ideal but may be impractical depending on the size of the site and scale of infestation. More weed-inundated sites may have to use larger scale techniques like solarization or smothering.

Solarization involves tilling or turning the soil, watering it thoroughly, and then covering it with clear plastic. It is important to completely seal the plastic by burying the perimeter. This, combined with the humidity from the water in the soil, will create a greenhouse effect when the sun hits it that sterilizes the seed bank in the soil and cooks any living plant material. This process may take a month or more depending on the amount of sun hitting the area. Simply covering an area with plastic, or using black plastic, will merely kill the weeds that were growing rather than destroy the seed bank, leading to another flush of weeds germinating once the plastic is removed.

Smothering is bringing in new soil that is not weed infested and burying the existing weeds and seed bank. At least 8 inches of material will be needed to suppress

One of the most important parts of site preparation is removing or suppressing nonnative weeds.

Many native plants, like this blanketflower (*Gaillardia aristata*), need good drainage.

tenacious, rhizomatous weeds. Putting down a thick layer of newspaper, cardboard, or weed cloth before bringing in the new material will aid in suppressing weeds and the seed bank, but this may have an undesirable effect on soil hydrology. Smothering does not destroy the existing seed bank. When digging holes for planting you will inevitably bring some of the preexisting soil to the surface along with the weed seeds in it so following up with hand pulling will always be needed.

Only use herbicides as a last resort and if you do only use ones with low residual action, meaning they will break down and become inert quickly. Make sure not to plant too soon after application and prevent pets and wildlife from entering the area. Follow all instructions and safety guidelines on the product label. Herbicides, except those with high residual action (which are not at all recommended), will not impact the existing seed bank so any time the soil is disturbed a new batch of weed seeds will be exposed and repopulate the area if not properly suppressed or managed. Herbicides should only be used (again, as a last resort!) to prepare a site for planting and never for managing weeds as the planting is establishing.

Planting and Mulching

Fall is a great time to plant native plants, especially in sites where irrigation for establishing plants is limited or unavailable. Planting in the fall gives plants time during the wet winter to settle in and begin to establish roots. Planting in late winter and spring is fine too, but plantings will require a bit more water to establish during the first year. Avoid planting in hot summer months or any month without rainfall unless you are able to irrigate. Most seeds of Pacific Northwest native plants need to be planted in fall so they can go through a cold, wet stratification period over the winter to break dormancy, allowing germination in spring. Seeds should not be planted any deeper than the seed is long; if broadcasted it helps to lightly rake seeds into the soil. Protect seeded areas from foraging birds and animals

Fall is a great time to plant native plants!

using Reemay, chicken wire, or a light shade cloth that is removed as seedlings germinate and begin to grow. Mark where you have planted seeds in case they take more than one year to germinate.

Plan for the future by properly spacing plants and selecting plants that grow well together. Planting species that naturally grow together in the wild will ensure they can cooperate, create functional habitat, and require the same watering regime in the future. Plan for layers in the garden by planting trees and shrubs along with low-growing perennials and groundcovers. To help give you some extra time to eliminate weeds from the planting site, consider establishing tree and shrub layers first, mulching heavily and weeding around them as they establish, later pulling heavy mulch aside and planting more delicate forbs and grasses using lighter forms or applications of mulch that will not choke out understory plants. Trees and shrubs may be small when first planted but will need ample space in the future. Spacing needs vary by species. Shrubs will need space to accommodate their mature width. Trees have roots that penetrate deeply into the soil and the surrounding area and should be given ample space from each other, as well as buildings, pipes, and pools. Denser spacing is desirable if trying to establish a hedge or screen.

Be sure to properly site plants based on the sunlight exposure they will receive. The time of day that plants receive sunlight is a factor to consider. Shade-loving plants may be able to handle morning or late-day sun but need to be protected from midday sun when the light and heat is more intense. Full sun exposure in places that tend to be cool and cloudy differs from that in hot and sunny areas. Providing shade for part of the day may be necessary even for sun-loving plants in extremely hot and sunny sites. Plants have optimal growing conditions, yet certain species may be able to survive outside of their ideal. In general, plants that can tolerate varying degrees of sun and shade tend to flower more profusely and have a more compact habit in sunnier conditions. In shadier sites the same plant may produce more vegetative growth but fewer flowers.

When planting, dig holes at least twice the width of the root ball or the container plants come in, and twice as deep if possible. Keep plants moist after you bring them home from the nursery until planting. Loosen roots before planting and water deeply and thoroughly after planting. Be careful not to bury plants with sensitive crowns too deeply in soil or with mulch as it may cause them to rot. Keep plants well watered for the first few weeks after planting and then appropriately watered as needed for establishment. Protect plants from pets and wildlife with caging or floating screen covers until properly established.

Mulching after planting helps to suppress weeds, retain soil moisture, improve soil health, and create an aesthetically pleasing look. Choose plants with various heights and growth habits so they eventually fill the space, reducing the

Plan for layers in the garden by planting trees and shrubs along with low-growing perennials and groundcovers.

Place plants properly based on site and light conditions.

Protect plants from pets and wildlife until properly established.

Broken-down leaves are often the ideal mulch, but some rock garden plants may respond well to a gravel mulch. Find the mulch that works best for the species in your garden.

need for mulch and producing their own when leaves fall. Coarser mulches can be used around trees and shrubs, especially while establishing to suppress weeds. Apply about 2 to 3 inches of mulch but be careful not to pile mulch too thickly against the trunks of trees and shrubs as it may encourage rot. Use finer mulches and thinner applications around grasses and forbs, particularly low-growing species or ones with sensitive crowns as mulch can smother those plants. Broken-down leaves are an ideal type of mulch and decompose easily, adding to and building soils. Different plants respond to different mulches. Species of cactus and other rock garden plants might thrive with a gravel mulch, trees may respond well to wood chips, and shady woodland plants love leaves and needles; find the mulch that works best for the species in your garden. "Bee" aware that thick applications and some types of mulch can negatively impact ground-nesting bees. Thin layers of pebbles or rocks, loose leaf litter, and leaving patches of bare ground are all beneficial for nesting bees.

Establishing and Watering Native Plants

Making sure plants become properly established is key to creating a native plant garden that can thrive in local climates and conditions, requiring less resources and continual inputs, especially water. To establish plants be prepared to irrigate, particularly during dry summer months, for the first year or two depending on the species and the site, possibly more for some trees and shrubs. Watering deeply and infrequently is much better than watering frequently and shallowly. Watering deeply encourages roots to grow down into the ground rather than staying close to the surface, which helps plants access more of the moisture held deep in the soil and eventually thrive without irrigation if properly sited. It is generally best to let soils dry out to about a finger's length deep before watering again. Many plants that grow in dry, sunny areas, such as desert parsleys (*Lomatium* spp.), circumvent summer drought by going dormant. These

Selecting species that are appropriate for your location and making sure plants become properly established is crucial to creating a native plant garden that can thrive in local climates with minimal irrigation.

Depending on the species, be prepared to irrigate plants for the first few years while they get established. Watering deeply and infrequently is better than watering frequently and shallowly.

plants may need to remain dry during summer dormancy and overwatering may cause them to rot and die. Be careful to water these plants very infrequently after they go dormant in summer and not at all after establishment. Do not plant such plants in areas that will be irrigated constantly and indefinitely. Bulbs typically take one year or less to establish and are best planted in fall. Most bulbs die back after blooming and may require soils to dry out after they go dormant. Gardeners may want to mark plants so they are not forgotten and accidentally disturbed when dormant. Have patience as some plants may remain small their first few years while establishing but will grow quickly once they gain a good foothold.

Except in years of severe drought, locally adapted and properly sited plants require minimal to no irrigation once established. Irrigation needs for species planted outside of their normal range may differ. Be prepared to celebrate plants in all stages of their yearly growth and anticipate that toward the end of summer parts of the garden may turn golden. Plants that are stress deciduous in seasonally dry conditions may be kept looking lush in summer with some supplemental water, and many species grow more robustly with occasional, deep irrigation in summer, but be careful not to water plants that require dry summers and are averse to summer irrigation. Don't forget that watering deeply and infrequently is best and keep in mind any plant in a container will need to be watered no matter how drought tolerant it is.

Maintaining a Native Plant Landscape

If you have chosen plants appropriate for your area and sited them properly in the garden according to their needs, carefully prepared your planting site, planted and mulched new plants, and given plants the water and care needed to

Watering, weeding, trimming growth, reapplying mulch, and continually providing for beneficial insects, pollinators, and wildlife are all a part of maintaining a thriving native plant garden.

Wild blue flax (*Linum lewisii*) spreads easily by seed.

become established, then you should be the proud steward of a thriving native plant garden. Maintenance at this point boils down to watering, weeding, trimming plant growth as needed, reapplying mulch as needed, and continually providing for and encouraging beneficial insects, pollinators, and wildlife.

Weeding is an ongoing task. Dense, healthy native plant communities will hopefully be able to outcompete nonnative weeds over time, especially with a little help from gardeners removing weeds before they go to seed. Native plants themselves may need thinning as they produce viable seed, which may fall to create more plants. Some species, such as wild blue flax (*Linum lewisii*) and self-heal (*Prunella vulgaris*),

spread easily by seed in ideal conditions. If you want to control them, you will need to remove seed before it ripens and falls or pull seedlings where they are not wanted. If you are growing annuals, allow plants to reseed themselves. Introduce plants to other parts of the garden by harvesting seed and sowing it into new areas or share the seed with friends and neighbors.

As plants grow, they may eventually need pruning or cutting back. Try to maintain a balance between form and function in the garden; keep it pretty but productive. Trimming plant growth should be done at times that does not harm nesting or hibernating wildlife and pollinators like bees and butterflies. Leave nesting materials for birds and

Try to maintain a balance between form and function in the garden.

beneficial insects and leave plant materials like leaves as a natural mulch and soil amendment. Keeping the garden too tidy discourages beneficial insects and wildlife who help maintain a natural balance of "garden pests."

Continually encourage biodiversity in the garden and be ready for change in the future. A native plant garden is a dynamic place, and the composition of your garden will evolve as plants mature and conditions change. Be prepared to replace plants that have senesced and cultivate new seedlings that develop. Most importantly, enjoy watching the garden grow, species flourish, and the knowledge that you are a part of a collective effort to replant and reinvigorate a thriving ecosystem.

Enjoy watching the garden grow, species flourish, and the knowledge that you are a part of a collective effort to replant and reinvigorate a thriving ecosystem.

How to Use This Book

Use this book to get excited about gardening with native plants and show your friends and neighbors that native plants are beautiful!

This book is intended to give the reader a basic understanding of select Pacific Northwest native plants and native plant gardening. Plants are grouped by category: wildflowers, grasses and grasslike plants, ferns, shrubs, and trees. In addition to a general summary with descriptions and other information that gardeners might find interesting or important, each plant profile highlights four important factors to consider in native plant selection: habitat and range, seasonal interest, wildlife value, and cultivation requirements. To quickly reference potential plants for specific gardening goals and needs, see Plants for Specific Purposes.

The Pacific Northwest has no official boundaries and could be considered as extensive as reaching from Southeast Alaska to Northern California and east to parts of the Rocky Mountains. This book primarily focuses on Washington and Oregon but may be applied to other areas in the region.

The plants featured in this book were chosen based on their beauty, form, and ecological function in the garden. We have selected species that give gardeners in every ecoregion something from which to choose. With some exceptions, the plants are generally abundant across the region or habitats in which they are found. Nursery availability was also taken into consideration so gardeners can easily and ethically source plants; however, availability of certain species may be limited.

The habitat and range information is important to consider regarding the plants you would like to incorporate into your garden. Selecting species that occur in your site's ecoregion and habitat type ensures the most success in healthy plant development and supporting local biodiversity and

Glacier fleabane (*Erigeron glacialis*)

ecological function. Alternate species for other ranges and habitats are noted in many plant profiles. More detailed information and maps of native plant distribution can be found online. See References and Resources for a list of recommended websites.

Seasonal Interest is highlighted for each plant so you can plan for continuous blooms, summer fruits, fall foliage, ephemeral emerald carpets of growth, and other features of a beautiful and functional native plant garden. By planting an assortment of species that bloom and provide interest at different times of the year you can ensure perpetual beauty in the garden and a sustained supply of nectar and forage for pollinators and wildlife.

The wildlife value section gives a simple synopsis of each plant's benefits to wildlife, although every plant in this book has ecological relationships that far exceed what we are able to present here. Deer resistance or desirability is noted when possible.

In addition, symbols can be found to the right of plant names to indicate which groups of wildlife—birds 🐦, hummingbirds 🦅, bees 🐝, butterflies 🦋, and/or caterpillars 🐛—directly benefit by using the plant as a food source, for shelter, or for nesting materials.

The cultivation requirements section explains the conditions, such as sunlight, soil, moisture, and maintenance plants need in order to thrive.

This book is not intended as a field identification tool. Native plant identification in the Pacific Northwest can be complex. There is a high amount of variation among some species and many sensitive, rare, and endemic plants may be hard to decipher. Always utilize proper identification and thorough research when working with plants in the wild.

We hope this book either enhances your existing ecological gardening library or is a jumping-off point for further research into the beauty, importance, and intricacies of the region's flora. There is no end to the knowledge that can be learned so we encourage you to explore the References and Resources provided at the end of this book to continue your personal education.

Learn more about native plants by joining your state's Native Plant Society. These organizations sponsor a wide variety of field trips and classes.

Wildflowers

Achillea millefolium • Asteraceae

Common yarrow

HABITAT/RANGE Widespread and adapted to most habitats. Alaska to California and east to the Atlantic Coast at low to high elevations. Circumboreal. Grows on both sides of the Cascade Mountains.

SEASONAL INTEREST White, occasionally pink, flowers spring to late summer or fall depending on location.

WILDLIFE VALUE Attracts butterflies, bees, and other pollinators. Butterfly and moth host plant. Attracts beneficial insects. Many solitary bees specialize on plants in the family Asteraceae. Provides food and nesting materials for birds. Typically deer resistant.

CULTIVATION Sun to part shade. Thrives in a variety of soil types, preferring ones that are not consistently saturated. Water to establish. Drought tolerant once established but some supplemental water will keep it looking lush in hot, dry sites. Deadhead to encourage additional blooms. Easy from seed. Can spread vigorously by seed and rhizomes; remove seed heads to help control spreading if needed. Does well in containers.

A familiar plant to gardeners, common yarrow is an attractive, aromatic perennial that grows up to 3 ft. tall with large clusters of small white flowers and fernlike foliage. It is fire resistant and tolerates mowing, which causes it to form a carpet of fragrant foliage, making it useful as a lawn alternative. A long-blooming pollinator favorite great for butterfly gardens and attracting beneficial insects, it's also useful for rehabilitating disturbed sites. There are many cultivars of yarrow available; choose plants wisely. Wild populations are highly variable and both native and introduced plants are found in the region.

Common yarrow has showy flat-topped clusters of flowers loved by butterflies and other pollinators.

Achlys spp. • Berberidaceae

Vanillaleaf

HABITAT/RANGE Moist conifer forests from southwestern British Columbia to northwestern California. Grows from the east base of the Cascade Mountains to the Pacific Coast at low to mid elevations.

SEASONAL INTEREST White flowers in spring to midsummer relative to elevation. Seasonal groundcover.

WILDLIFE VALUE Wind pollinated yet still attracts insect pollinators. Moth host plant. Provides cover for ground-nesting birds and small animals.

CULTIVATION Full to part shade and moist to seasonally moist, well-drained, humus-rich soil. This understory plant requires ample amounts of organic matter in the soil. Water to establish. Tolerates seasonally dry conditions once established if properly sited but providing some supplemental water in summer will keep plants looking lush. Can grow under conifers. Slow and sometimes difficult to establish. Spreads slowly by rhizomes. A lovely groundcover for moist woodland gardens. Mulch.

Vanillaleaf is an attractive understory plant that creates a verdant groundcover in moist forests. From its spreading rhizomes this herbaceous perennial produces large leaves and spikes of small white flowers that grow about 1 ft. tall. The leaves are divided into three prominently lobed, winglike leaflets that have a sweet fragrance when dried. There are two genetically distinct but visually similar species of vanillaleaf found in the Pacific Northwest, *Achlys triphylla* and *A. californica*. They share the same range and habitat, and mixed populations are common. Both are a good choice for gardens west of the Cascades.

Vanillaleaf is a lovely groundcover for moist, shady forest gardens.

Actaea rubra • Ranunculaceae

Red baneberry

HABITAT/RANGE Moist woodlands, riparian areas, and coastal areas. Alaska to California and east to the northern Atlantic Coast at low to high elevations. Grows on both sides of the Cascade Mountains.

SEASONAL INTEREST White flowers in spring to midsummer relative to elevation. Ornamental red or white berries in summer.

WILDLIFE VALUE Poisonous to humans but not most wildlife. Fruits are eaten by birds and small mammals. Attracts pollinators. Infrequently browsed by deer and elk.

CULTIVATION Part to full shade and moist, humus-rich soil. Tolerates more sun in cooler locations. Water to establish. Tolerates seasonally dry conditions once established if properly sited. Grows well under conifers. Plant in moist, shady gardens. Mulch.

Red baneberry's clusters of white, rose-scented flowers brighten dark, shady woodlands in spring, but the real attraction in the garden is its ornamental red or white berries. These berries are beautiful but very poisonous and gardeners with young children are cautioned not to plant *Actaea rubra*. This herbaceous perennial can grow 1 to 3 ft. tall, has attractive foliage with sharply toothed leaves, and looks lovely interspersed with evergreen ferns in moist woodland gardens.

Actaea rubra produces white flowers and ornamental red or white berries.

Agastache urticifolia • Lamiaceae

Nettleleaf horsemint

HABITAT/RANGE Grows in a variety of habitats including meadows, open woodlands, and slopes. Southern British Columbia to California and east to the Rocky Mountains, from foothills to fairly high elevations. Grows mainly east of the Cascade Mountains.

SEASONAL INTEREST Pinkish purple and white flowers in late spring to summer relative to elevation.

WILDLIFE VALUE An important plant for pollinators. Favored by butterflies, hummingbirds, hummingbird moths, and bees. Monarch butterfly nectar plant. Seeds eaten by birds. Typically ignored but may be browsed by deer.

CULTIVATION Sun to part shade and moist to seasonally dry, well-drained soils. Water to establish. Fairly drought tolerant once established but benefits from some supplemental water. Goes dormant after setting seed in drier sites. Mulch lightly.

An aromatic powerhouse of a plant for pollinators and one of the best butterfly plants in the region, *Agastache urticifolia* has dense, showy spikes of pinkish purple and white flowers on stems that grow around 3 ft. tall or more from woody crowns. Although this herbaceous perennial is adaptable and can grow in seemingly hot and dry habitats, it prefers some moisture in the soil. An attractive plant for the butterfly garden, meadowscape, or open woodlands, it also makes a lovely cut flower that dries well and has a minty fragrance. For cut flowers, take from the garden only, not from the wild.

Agastache urticifolia has a minty fragrance and showy flowers that are loved by butterflies.

Ageratina occidentalis • Asteraceae

Western snakeroot

HABITAT/RANGE Rocky places and gravelly streambanks. Central Washington to California and east to Montana at low to high elevations. Listed as sensitive and at risk in Montana. Grows east of the Cascade Mountains in Washington but on both sides of the Cascades from Oregon south.

SEASONAL INTEREST Pink to purple or white flowers in summer.

WILDLIFE VALUE Attracts bees, butterflies, and other pollinators. Many species of solitary bees specialize on plants in the family Asteraceae.

CULTIVATION Sun to part shade and moist to seasonally dry, well-drained soil. Water to establish. Tolerates seasonally dry conditions once established if properly sited. Grows and flowers best with occasional summer water, especially in hot, dry areas. Loves to grow in rocky soils and is a good choice for rock gardens. Grows well in containers. Mulch lightly.

Western snakeroot is a lovely deciduous perennial with triangular, toothed leaves that can grow to 2 ft. tall from a woody crown. Although rhizomatous, it generally maintains a clumping habit and loves to grow among rocks. It is a great plant for butterfly gardens and its flowers are favored by many kinds of pollinators. As its Latin name implies, its fuzzy looking, pink to purple or white flowers are reminiscent of *Ageratum*, a commonly cultivated genus of plants native to warm climates. This plant was classified as *Eupatorium occidentale* and is also called western boneset or western Joe Pye weed.

Western snakeroot has fuzzy looking flowers that resemble ageratum.

Allium acuminatum • Amaryllidaceae

Tapertip onion

HABITAT/RANGE Open, dry, rocky hillsides, meadows, grasslands, shrub-steppe, and woodland edges. Southern British Columbia to California and east to the Rocky Mountains at low to mid elevations. Grows on both sides of the Cascade Mountains.

SEASONAL INTEREST Deep pink to light rose-purple or white flowers in late spring to midsummer.

WILDLIFE VALUE Attracts butterflies, bees, hummingbirds, and other pollinators. Attracts beneficial insects. Bulbs eaten by mammals. Typically ignored but may be browsed by deer.

CULTIVATION Sun to light shade and rocky, well-drained soil. Prefers soils that are moist in spring but dry in summer. Drought tolerant once established. Water to establish but water infrequently, if at all, after it finishes setting seed and goes dormant. A good plant for rock gardens, parking strips, and dry sunny meadows.

Tapertip onion is a small perennial bulb with grasslike leaves and umbels of pink, sometimes white, flowers on short, leafless stalks that grow up to 1 ft. tall. The entire plant has an oniony aroma and can be eaten in small amounts. There are many wild onions native to the Pacific Northwest and *Allium acuminatum* is one of the most common. It provides a drought-tolerant, pollinator-friendly option for gardeners, especially east of the Cascades. Slim-leaf onion (*A. amplectens*) also grows in dry, rocky sites that are vernally moist and is more abundant west of the Cascades, particularly in Oregon and California.

Allium acuminatum is a small onion with showy flowers that grows in dry, rocky sites.

Allium cernuum • Amaryllidaceae

Nodding onion

HABITAT/RANGE Moist to vernally moist meadows, hillsides, rocky areas, and open woodlands in both coastal and mountainous areas. British Columbia to northern Oregon and east to the Rocky Mountains and other areas of the United States. Low to high elevations. Grows on both sides of the Cascade Mountains.

SEASONAL INTEREST Pink, sometimes white, flowers in late spring to summer.

WILDLIFE VALUE Attracts butterflies, bees, hummingbirds, and other pollinators. Butterfly host plant. Attracts beneficial insects. Bulbs eaten by mammals. Typically ignored but may occasionally be browsed by deer.

CULTIVATION Sun to part shade and moist to vernally moist, well-drained soil. Can grow in rocky, sandy, or clay soils. Provide some shade in hotter sites. Water to establish. Drought tolerant once established if properly sited. Increases by bulb offsets forming clumps, which can be divided. Spreads vigorously by seed in ideal conditions although may not compete well against heavy vegetation. Collect seed before it falls to curtail spreading if needed. A good plant for rock gardens and edible landscapes. Grows well in containers.

Allium cernuum is a perennial bulb with grasslike leaves and nodding clusters of pink, sometimes white, bell-shaped flowers. Growing 6 to 20 in. tall, the entire plant has a mild onion scent and can be eaten in small amounts. It makes a lovely garden addition and can grow in moist to seasonally dry soils. There are many species of wild onion native to the Pacific Northwest. Pacific onion (*A. validum*) grows in moist areas and ranges farther south than *A. cernuum*.

Allium cernuum **forms clumps and has lovely nodding flowers.**

Anaphalis margaritacea • Asteraceae

Pearly everlasting

HABITAT/RANGE Open forests, rocky slopes, meadows, and roadsides. Alaska to California and across North America at low to high elevations. Widespread. Grows on both sides of the Cascade Mountains.

SEASONAL INTEREST Blooms summer to early fall with white, papery bracts that persist through fall.

WILDLIFE VALUE Attracts butterflies and a wide variety of pollinating and beneficial insects. A great plant for small bees. Larval host for the painted and American lady butterfly. Deer resistant.

CULTIVATION Sun to part shade and moist to seasonally dry, well-drained soil. The more sun this plant gets the more moisture it may need. Water to establish. Drought tolerant once established. Requires good drainage but can grow in a variety of soil types. Does well in rocky soils. Spreads vigorously by rhizomes. Grows well in planters but will outcompete other plants. Mulch.

Pearly everlasting has many attractive qualities, top among them is its ability to provide nectar to pollinators and interest in the garden late in the growing season. This rhizomatous herbaceous perennial grows 1 to 3 ft. tall with leafy, white-woolly stems topped by showy clusters of flowers. The small yellow flowers are surrounded by white, papery bracts that persist long after flowering and dry well, making this a favored plant for bouquets. Drawing in clouds of small pollinators such as skipper butterflies and small native bees when in bloom, pearly everlasting is an excellent choice for butterfly, pollinator, and xeric gardens.

The white, papery bracts on the inflorescences of pearly everlasting create long-lasting interest.

Anemone deltoidea • Ranunculaceae

Columbia windflower

HABITAT/RANGE Moist to dry forests from Puget Sound to Northern California. Grows at low to mid elevations mainly in and west of the Cascade Mountains, farther east in the Columbia River Gorge.

SEASONAL INTEREST White flowers in spring to summer depending on location.

WILDLIFE VALUE Attracts pollinators.

CULTIVATION Full to part shade and moist, humus-rich soil. A woodland plant that grows well under trees and shrubs. Water to establish. Tolerates seasonally dry conditions once established if properly sited. Some supplemental water in summer will keep plants looking lush longer. Mulch lightly.

A lovely low-growing plant for woodland gardens and shady sites. *Anemone deltoidea* is an herbaceous perennial that spreads unaggressively by slender rhizomes to form colonies. Solitary stems support a single showy white flower and grow 4 to 12 in. tall with a whorl of three leaves at midstem. There are a few species of *Anemone* native to the region that can be charming garden plants. Cliff anemone (*A. multifida*) is easy to grow and has a lovely clumping habit and foliage, along with white to pink flowers that bloom in spring and sometimes again in fall. Its attractive seeds are densely covered in soft hairs and collected by hummingbirds for nesting material. Western pasqueflower (*A. occidentalis*), which grows at mid to high elevations, and Oregon anemone (*A. oregana*), a blue-flowered woodland plant, are interesting and attractive but may be more difficult to cultivate.

Anemone deltoidea brightens moist, shady woodlands in spring with brilliant white blooms.

Angelica lucida • Apiaceae

Sea-watch

HABITAT/RANGE Coastal meadows, marshes, beaches, headlands, and streambanks. A strictly coastal species in Oregon, Washington, and Northern California that ranges farther inland in northern British Columbia and Alaska.

SEASONAL INTEREST White flowers late spring to summer. Tall seed stalks have lasting interest.

WILDLIFE VALUE Attracts butterflies, bees, and other pollinators. Host plant for the anise swallowtail butterfly. Provides cover for wildlife.

CULTIVATION Sun to part shade and moist soils. Can grow in salty and sandy soils. Water to establish. No supplemental water needed after establishment in coastal areas; provide supplemental water elsewhere. Give this large species plenty of space. Mulch.

Sea-watch is a robust herbaceous perennial that grows 2 to 5 ft. tall with large compound leaves and dense umbels of white flowers. This plant is attractive in flower and in fruit and provides structural interest to coastal butterfly gardens. There are other species of *Angelica* native to the Pacific Northwest that occur inland and in more varying habitats than *A. lucida*, all of which create striking garden accents. Sharptooth angelica (*A. arguta*) blooms in summer and grows to 6 ft. tall in meadows and wetlands at low to mid elevations on both sides of the Cascade Mountains. Kneeling angelica (*A. genuflexa*) grows 3 to 5 ft. tall in moist places mainly west of the Cascades.

Angelica lucida is a coastal species whose large seed heads create lasting interest.

Aquilegia formosa • Ranunculaceae

Red columbine

HABITAT/RANGE Grows in a variety of habitats including forests, moist meadows, rocky slopes, coastal areas, and riparian areas. Alaska to California and east to the Rocky Mountains at low to high elevations. Widespread. Grows on both sides of the Cascade Mountains.

SEASONAL INTEREST Red and yellow flowers in spring to summer relative to elevation.

WILDLIFE VALUE A magnet for hummingbirds. Attracts bees, bumble bees, butterflies, and other pollinators. Moth host plant. Seeds eaten by birds. Deer resistant.

CULTIVATION Part shade and moist to seasonally moist, well-drained soil. Will grow in full sun if there is adequate moisture available. Does well in humus-rich soil but tolerates lean soils as well. Water to establish. Drought tolerant once established if properly sited. Goes dormant by mid- to late summer in hotter, drier sites without supplemental water. Sometimes a short-lived perennial but typically reseeds readily. Seed can take two years to germinate. Protect from slugs. Mulch lightly.

Red columbine is one of the region's most beautiful and intriguing wildflowers. It grows 1 to 3 ft. tall with many nodding flowers consisting of petals that form red, nectar-filled spurs above and yellow blades below flared red sepals. The curious flower shape is an evolutionary treat designed for hummingbirds and this popular perennial is a must-have for hummingbird and pollinator gardens. Columbines are promiscuous and will easily hybridize with both wild and cultivated relatives. Yellow columbine (*Aquilegia flavescens*) grows in mountainous areas east of the Cascades.

Red columbine's unique flowers are favored by hummingbirds.

Armeria maritima • Plumbaginaceae

Sea-pink

HABITAT/RANGE Coastal beaches and bluffs, occasionally along streams and in prairies near the coast from Alaska to California. One subspecies also grows inland in Alaska. Circumboreal.

SEASONAL INTEREST Pink to purplish flowers in spring. Sporadic flowering may occur throughout the summer into fall. Evergreen foliage.

WILDLIFE VALUE Attracts bees, butterflies, hummingbirds, and other pollinators. Attracts beneficial insects. Deer resistant.

CULTIVATION Sun and well-drained, infertile soils. Does well in sandy and rocky soils. Will rot out in saturated, overly rich soils. Water to establish. Drought tolerant once established if properly sited, especially in coastal areas. May benefit from some shade or a cool spot and supplemental water if planted inland. Use in rock gardens and along garden edges. Grows well in containers.

If the tidy mats of evergreen, needlelike leaves of *Armeria maritima* are not reason enough to make it an excellent garden selection, its profusions of bright pink to purplish flowers, which may bloom multiple times throughout the year, make it completely irresistible. The clusters of flowers bloom on top of leafless stalks that grow up to 1 ft. tall. Multiple subspecies occur in North America. In our area subsp. *californica* is native but subsp. *maritima*, introduced from Europe, occurs sporadically from Victoria to Yaquina Head in Oregon. Many cultivars and selections of this popular perennial are available in nurseries; always favor planting true natives.

Sea-pink is a low-growing, mat-forming perennial that produces showy, pink clusters of flowers.

Arnica cordifolia • Asteraceae

Heartleaf arnica

HABITAT/RANGE Open woodlands, coniferous forests, and subalpine meadows. Widespread from southern Alaska to California and east to the Rocky Mountains and Great Lakes at low to high elevations. Grows on both sides of the Cascade Mountains.

SEASONAL INTEREST Yellow flowers in spring to summer depending on location. Seasonal groundcover.

WILDLIFE VALUE Attracts bees, butterflies, moths, and other pollinators. Many solitary bees specialize on plants in the family Asteraceae. May be browsed by deer and elk.

CULTIVATION Part shade and moist to seasonally dry, humus-rich soils. Water to establish. Drought tolerant once established if properly sited. Goes dormant by midsummer in drier sites. Occasional supplemental summer water may keep this plant green and produce a second bloom but stop watering when it dies back. Spreads unaggressively by rhizomes to form low-growing colonies. Grows well under conifers but if the canopy becomes too dense it will diminish, returning abundantly after a disturbance allows in light. Mulch lightly.

Heartleaf arnica's cheerful yellow flowers brighten forest openings and understories, growing 4 to 24 in. tall from rhizomes that can survive drought and low-intensity fires. Its flowering stems, which produce one to three flowers, and nonflowering shoots bear heart-shaped basal leaves. This is a lovely pollinator-friendly seasonal groundcover for open, eastside forest gardens and partly shaded xeric gardens. There are many species of *Arnica* native to the Pacific Northwest. Some species, like *A. cordifolia*, are drought tolerant while others require moist soils.

Arnica cordifolia brings a bit of sunshine to the understory of dry, open woodlands.

Artemisia ludoviciana • Asteraceae

Western mugwort

HABITAT/RANGE Meadows, open slopes, shrub-steppe, grasslands, and riparian areas. This species, with its many subspecies, occurs in various habitats across North America. Low to high elevations. Grows mainly east of the Cascade Mountains.

SEASONAL INTEREST Silvery foliage.

WILDLIFE VALUE Wind pollinated. Butterfly and moth larval host. An excellent plant for attracting beneficial insects. Provides food and cover for birds. Forage for antelope and elk. Deer resistant.

CULTIVATION Sun to light shade and well-drained soils. Water to establish. Drought tolerant once established. Some subspecies spread aggressively by rhizomes in moist soils; drier soils will limit growth. Leave stalks over winter for wildlife habitat. Trim back old growth in spring when new growth emerges. Does well in planters though it will dominate the container. Mulch.

Western mugwort, also called white sagebrush or silver wormwood, is an aromatic perennial that grows up to 3 ft. tall. It has rather inconspicuous, yellow, wind-pollinated flowers, but is a striking plant for sunny gardens due to its silvery foliage, which can provide height and contrast. It is useful for attracting beneficial insects to farms and gardens and can be used in bouquets. This species is comprised of several subspecies that vary in growth habit as well as preferred habitat, but typically this is a drought-tolerant, sun-loving species that is easy to grow.

The silvery, aromatic foliage of western mugwort provides contrast in the garden.

Aruncus dioicus var. *acuminatus* • Rosaceae

Goatsbeard

HABITAT/RANGE Moist forests, wetlands, coastal areas, riparian areas, and mountainous areas. Alaska to California. Low to high elevations. Grows primarily west of the crest of the Cascade Mountains.

SEASONAL INTEREST White flowers in spring to summer depending on location. Some fall color.

WILDLIFE VALUE Attracts beneficial insects including braconid wasps. Attracts bees, butterflies, hummingbirds, and other pollinators. Host plant for the dusky azure butterfly. Seeds eaten by birds. Provides cover. Browsed by deer and elk.

CULTIVATION Full to part shade and moist, humus-rich soil. Tolerates more sun in cool, wet locations. Water to establish and continue to provide supplemental water as needed. Spreads slowly by short rhizomes. Use in rain gardens, moist woodlands, and riparian areas. Grows well in containers. Mulch.

Goatsbeard is a graceful, robust, shade-loving perennial that can grow 3 to 6 ft. tall and 3 to 5 ft. wide with showy plumes of small white flowers and large compound leaves comprised of many toothed leaflets. Plants are either male or female, and though nearly similar, male flowers are somewhat showier. *Aruncus dioicus* var. *acuminatus* makes a bold accent and is a good background plant because of its height and size. Give it ample space. This is a popular landscaping plant for moist, shady gardens and there are various cultivars available in nurseries, as well as selections from species native to Asia and Europe. Select plants wisely and favor planting true, locally appropriate natives.

Goatsbeard is a large and bold plant for moist, shady gardens.

Asarum caudatum • Aristolochiaceae

Wild ginger

HABITAT/RANGE Moist forests at low to mid elevations from British Columbia to California and east to northern Idaho and Montana. Grows primarily in and west of the Cascade Mountains.

SEASONAL INTEREST Evergreen groundcover across most of its range. Maroon flowers, occasionally greenish white, spring to midsummer.

WILDLIFE VALUE Flowers attract pollinating insects but are also self-pollinating. Seeds dispersed by ants. Sometimes browsed by small mammals. Typically ignored but occasionally browsed by deer.

CULTIVATION Full to part shade and moist, humus-rich soil. Water to establish. Tolerates seasonally dry conditions once established if properly sited. Supplemental water will be needed in drier sites. Spreads slowly by rhizomes to form large mats. Use as a groundcover under trees and along the edges of moist, shady gardens. Protect young plants from slugs. Mulch but don't bury deeply.

With thick, fleshy rhizomes this low-growing perennial creeps slowly through moist forests of the Pacific Northwest. Beneath a dense layer of attractive, leathery, heart-shaped leaves that remain evergreen in mild winters, wild ginger hides strange maroon, sometimes greenish white, flowers. The entire plant has a distinct aroma of ginger when crushed, though it is not related to culinary ginger. This is one of the region's best groundcovers for moist, shady sites.

Wild ginger is a lovely aromatic groundcover.

Asclepias fascicularis • Apocynaceae

Narrow-leaved milkweed

HABITAT/RANGE Meadows, streambanks, open areas, and disturbed sites. Northeastern Washington to Mexico and east to Idaho at low to mid elevations. Grows on both sides of the Cascade Mountains, but only east of the Cascades in Washington.

SEASONAL INTEREST Pinkish purple flowers in early summer to early fall depending on location. Long pods with silky haired seeds.

WILDLIFE VALUE Larval host for the monarch butterfly. Important for pollinators. Attracts bees, butterflies, hummingbirds, and other pollinators. Attracts beneficial insects. Birds glean insects and use fibers from the plant for nesting material. Deer resistant.

CULTIVATION Sun and moist to seasonally dry, well-drained soils. Tolerates clay. Water to establish. Drought tolerant once established if properly sited. Spreads by rhizomes. Often hosts oleander aphids, which usually only feed on plants in the family Apocynaceae. Do not use pesticides on this plant; it will attract the beneficial insects needed to keep aphid populations in check. Mulch lightly.

Narrow-leaved milkweed is an attractive herbaceous perennial that grows 1 to 3 ft. tall and provides nectar to a wide range of pollinators at a time of year when many other drought-tolerant plants have finished blooming. Its curiously structured, deep to light pinkish purple flowers are showy and bring summer color to sunny xeric gardens. An important butterfly host plant and a great choice for pollinator gardens and beneficial insect plantings. More drought tolerant than showy milkweed (*Asclepias speciosa*).

Narrow-leaved milkweed is a beautiful, drought-tolerant perennial that attracts a variety of pollinators.

Asclepias speciosa • Apocynaceae

Showy milkweed

HABITAT/RANGE Meadows, grasslands, riparian areas, roadsides, and ditches. British Columbia to California and east to the Mississippi Valley. Mostly at low elevations. Grows on both sides of the Cascade Mountains, but only east of the Cascades in Washington.

SEASONAL INTEREST Pink to purplish flowers late spring to midsummer. Large seed pods.

WILDLIFE VALUE Larval host for the monarch butterfly. Important for pollinators. Attracts bees, butterflies, hummingbirds, and other pollinators. Attracts beneficial insects. Birds glean insects from plants. Deer resistant.

CULTIVATION Sun and moist to seasonally moist soil. Prefers loamy to sandy soils but tolerates clay. Water to establish. Tolerates seasonally dry conditions but prefers moderate soil moisture. Spreads aggressively by rhizomes. Can be grown in containers. Often hosts oleander aphids, which usually only feed on plants in the family Apocynaceae. Do not use pesticides on this plant; it will attract the beneficial insects needed to keep aphid populations in check. Mulch.

Showy milkweed is one of the most wonderfully fragrant plants in the region. The large clusters of pink to purplish flowers are richly perfumed and flamboyantly ornate. Its large seed pods and silky haired seeds also bring interest to the garden. This is a robust perennial that grows up to 4 ft. tall and spreads by stout, fast-growing rhizomes. Larval host for the monarch butterfly and one of the best pollinator plants in the region, showy milkweed is a must-have for sunny butterfly and pollinator gardens.

Showy milkweed produces clusters of incredibly fragrant flowers.

Balsamorhiza spp. • Asteraceae

Balsamroot

HABITAT/RANGE Dry slopes, meadows, grasslands, oak woodlands, shrub-steppe, and forest edges and openings. There are many species of balsamroot, all native to western North America. Generally at low to mid elevations. Grows on both sides of the Cascade Mountains, though most species occur to the east.

SEASONAL INTEREST Yellow flowers in spring to early summer.

WILDLIFE VALUE Important for pollinators and wildlife. Attracts butterflies, bees, and many other pollinating and beneficial insects. Moth host plant. Seeds favored by birds and mammals. Provides cover. Browsed by deer and elk.

CULTIVATION Sun to light shade and seasonally dry, well-drained soils. Water to establish, but infrequently, if at all, when dormant in summer. Drought tolerant once established. Do not plant in irrigated beds. Protect sown seed from predation. Takes many years to develop and flower but is worth the wait. Goes dormant after setting seed. Leave plant material as cover for insects.

Balsamroots are beautiful herbaceous perennials that can grow 1 to 3 ft. tall and wide with taproots 5 to 9 ft. deep. Arrowleaf balsamroot (*Balsamorhiza sagittata*) and Carey's balsamroot (*B. careyana*) grow well in gardens east of the Cascades. Deltoid balsamroot (*B. deltoidea*) is the best choice for gardeners west of the Cascades. These are all robust species that benefit pollinators and wildlife and look stunning in dry, sunny gardens. Although balsamroot can appear abundant, some species are rare in parts of their range. Hybridization makes identification difficult in some areas. Never attempt to dig these deeply taprooted plants from the wild.

Balsamroot is an iconic western wildflower with sunflower-like blooms.

Camassia leichtlinii • Asparagaceae

Great camas

HABITAT/RANGE Vernally moist meadows, prairies, slopes, riparian woodlands, and oak savannas. Southern British Columbia to California at low to mid elevations. Grows mainly west of the Cascade Mountains.

SEASONAL INTEREST Deep blue-violet flowers in spring. One subspecies has creamy white flowers.

WILDLIFE VALUE Nectar plant for the endangered Fender's blue butterfly. Attracts butterflies, bees, hummingbirds, and other pollinators. Attracts beneficial insects. Bulbs eaten by mammals. Browsed by deer and elk.

CULTIVATION Sun to part shade and vernally moist sites that are wet in spring but dry in summer. Prefers humus-rich soils but tolerates clay. Water to establish and keep soils moist until plants finish blooming; stop watering or only water infrequently when plants go dormant. Goes dormant after setting seed. Intersperse with summer blooming plants. May spread vigorously by seed; remove seed before it falls to curtail spreading if needed. Can take three years to bloom from seed. Protect from slugs and gophers until established. Mulch.

Great camas is a perennial bulb with grasslike leaves and showy deep blue-violet flowers on sturdy stems that grow 2 to 3 ft. tall. It makes a colorful ephemeral statement in gardens and is a beautiful addition to wildflower meadows, as well as butterfly and pollinator plantings. A good choice for gardeners west of the Cascades.

Patches of great camas put on colorful displays in moist meadows in spring.

Camassia quamash • Asparagaceae

Common camas

HABITAT/RANGE Vernally moist meadows, prairies, oak savannas, open woodlands, and vernal pools. Grows from southern British Columbia to California and east to Montana at low to mid elevations. Found on both sides of the Cascade Mountains.

SEASONAL INTEREST Dark to light blue, sometimes white, flowers in spring.

WILDLIFE VALUE Nectar plant for the endangered Fender's blue butterfly. Attracts butterflies, bees, hummingbirds, and other pollinators. Attracts beneficial insects. Bulbs eaten by mammals. Browsed by deer and elk.

CULTIVATION Sun to light shade and vernally moist sites that are wet in spring but dry in summer. Prefers humus-rich soils but tolerates a variety of soils including clay. Water to establish and keep soils moist until plants finish blooming; stop watering or only water infrequently when plants go dormant. Goes dormant after setting seed. Can take three years to flower from seed. Protect from slugs and gophers. Mulch.

Common camas brings lavish amounts of color to seasonally moist meadows in spring but, sadly, abundant fields of camas are not as common as they once were. This lovely ephemeral has grasslike leaves and dark to light blue, sometimes white, flowers on stems that grow 1 to 2 ft. tall. Used ornamentally both in the Pacific Northwest and abroad, this perennial bulb makes a great addition to wildflower meadows, butterfly gardens, and rain gardens on either side of the Cascades. Several subspecies comprise this wide-ranging and variable wildflower. Plant starts ethically propagated from locally sourced seed.

Common camas is a beautiful butterfly plant and well-known wildflower.

Campanula rotundifolia • Campanulaceae

Bluebell bellflower

HABITAT/RANGE Rocky balds, cliffs, grassy slopes, meadows, forest openings, and gravelly streambanks from coastal to alpine areas. Widespread throughout most of North America. Circumboreal. Low to high elevations. Grows on both sides of the Cascade Mountains.

SEASONAL INTEREST Blue-violet flowers in summer to early fall.

WILDLIFE VALUE Attracts bees, butterflies, hummingbirds, and other pollinators. Specialist bee host. Deer resistant.

CULTIVATION Sun to part shade and well-drained soil. Water to establish. Drought tolerant once established if properly sited. Prefers part shade and occasional water in hot, dry areas. Deadhead to prolong bloom time. Basal leaves tend to wither when plants bloom. Spreads by slender rhizomes and can be used as a groundcover. May overcome other low-growing plants; give it space. Beautiful in planters and rock gardens. Too much mulch can smother this low-growing perennial.

Campanula rotundifolia is tougher and more drought tolerant than its delicate foliage makes it seem. Spreading gracefully, it forms a low-growing mat of rounded basal leaves below sprawling to erect, 1 to 2 ft. long flowering stems with linear leaves. Its blue-violet, bell-shaped flowers can bloom throughout summer, especially with a little deadheading and occasional water, providing long-lasting garden color and forage for pollinators. This pretty, pollinator-friendly plant makes a great groundcover in partly shaded, low-water gardens.

Campanula rotundifolia is a beautiful summer bloomer with attractive flowers and foliage.

Castilleja miniata • Orobanchaceae

Scarlet paintbrush

HABITAT/RANGE Moist mountain meadows, grassy slopes, coastal bluffs, and open forests. Widespread from Southeast Alaska to California and east to the Rocky Mountains and central Canada at low to mid elevations. Grows on both sides of the Cascade Mountains.

SEASONAL INTEREST Scarlet "flowers" late spring to early fall depending on location.

WILDLIFE VALUE Attracts hummingbirds and butterflies. Host plant for checkerspot butterflies.

CULTIVATION Sun to part shade and moist to seasonally moist soil. Water to establish. Tolerates seasonally dry conditions once established if properly sited. Often difficult to establish but well worth it if successful. Species of *Castilleja* are hemiparasitic, meaning they gain nutrients from the roots of nearby plants but are able to photosynthesize and produce nutrients on their own, though they are more likely to thrive with a host plant. They do not visibly impair their hosts. Plant starts or sow seed next to other perennials or grasses such as Oregon sunshine, yarrow, penstemons, asters, and fescue. Do not transplant.

It is not petals but rather leafy bracts that are bright red and make scarlet paintbrush a showy and colorful wildflower that attracts hummingbirds. Flower stalks grow 8 to 32 in. tall from woody crowns with green to burgundy foliage. Two variations of this perennial wildflower occur in our area, var. *dixonii* grows at low elevations along the coast. The Pacific Northwest is home to many species of *Castilleja* ranging in abundance from common to endangered. Harsh paintbrush (*Castilleja hispida*) is another attractive, widespread species that is drought tolerant.

Scarlet paintbrush is a colorful, iconic wildflower.

Chamaenerion angustifolium • Onagraceae

Fireweed

HABITAT/RANGE Open slopes, meadows, forest edges and clearings, streambanks, coastal areas, and disturbed areas; particularly prolific in burned sites. Widespread from Alaska to California and east to the Atlantic Coast at low to high elevations. Circumboreal. Grows on both sides of the Cascade Mountains.

SEASONAL INTEREST Deep pink flowers summer to early fall.

WILDLIFE VALUE Loved by pollinators including bees, butterflies, and hummingbirds. Moth host plant. Attracts beneficial insects. Nesting sites and nesting materials for cavity-nesting bees. Provides cover. Important food source, especially after a forest fire. Browsed by elk and deer.

CULTIVATION Sun to part shade and moist to seasonally moist soil. Fast-growing. Spreads by rhizomes. While it can be said "good fences make good neighbors" it could also be said "good fireweed placement makes good neighbors." This plant likes to travel and will eventually creep into new areas, making it unsuitable for small spaces. Water to establish and continue to provide supplemental water in dry sites; however, drier soils may restrict spreading. Leave stalks over winter for cavity-nesting bees. Great for wildflower meadows, pollinator plantings, and wilder parts of the garden. Mulch.

Fireweed is an attractive, vigorous herbaceous perennial that can grow quite tall, reaching as high as 8 to 9 ft. in ideal conditions. Racemes of deep pink flowers bloom continuously through summer, making it a pollinator favorite and a long-lasting color accent in the garden. Plant in areas where the creeping rhizomes are welcome to move freely around other plants. Honey bee keepers will want to plant fireweed as it is prized for honey production.

Fireweed is a tall, showy, long-blooming perennial loved by bees and hummingbirds.

Clarkia pulchella • Onagraceae

Elkhorns clarkia

HABITAT/RANGE Dry meadows, rocky slopes, grasslands, open forests, and shrub-steppe. From southern British Columbia to southeastern Oregon and east to western Montana at low to mid elevations. Grows mainly east of the Cascade Mountains.

SEASONAL INTEREST Rose to lavender flowers late spring to early summer.

WILDLIFE VALUE Attracts bees, butterflies, hummingbirds, and other pollinators. Specialist bee host.

CULTIVATION Sun to light shade and vernally moist soils with good drainage. Water to establish. Drought tolerant but some supplemental water may increase growth and prolong bloom time. Annual. If planting seed, be sure it is ethically sourced as seed collection of annual plants can have a big impact on wild parent populations. Allow garden plants to reseed. Plant among drought-tolerant perennials, keeping in mind that it does not compete well with heavy pressure from other vegetation. Good for xeric and rock gardens.

Elkhorns clarkia is a colorful, low-growing annual wildflower with branching stems that can grow 20 in. tall and short racemes of interesting flowers with four deeply lobed, rose- to lavender-colored petals. This plant is happiest growing east of the Cascades on dry, sunny slopes that are moist in spring. There are a few species of *Clarkia* native to the Pacific Northwest that make lovely annual garden plants and attract pollinators. Common clarkia (*C. rhomboidea*) also grows on dry slopes and open areas mainly east of the Cascades, while farewell-to-spring (*C. amoena*) is the best choice for gardeners west of the Cascades.

Elkhorns clarkia is a colorful and unique annual wildflower.

Claytonia perfoliata • Montiaceae

Miner's lettuce

HABITAT/RANGE Vernally moist areas, open woodlands, and disturbed sites. Southern British Columbia south through California and east to Montana at low to mid elevations. Grows on both sides of the Cascade Mountains.

SEASONAL INTEREST White flowers bloom in spring to early summer relative to elevation. Edible foliage. Seasonal groundcover.

WILDLIFE VALUE Attracts bees, butterflies, and other pollinators. Moth host plant. Browsed by small mammals and quail. Seeds eaten by birds.

CULTIVATION Part shade and soils that are moist in spring. Grows best in humus-rich soil but will tolerate a variety of soil types. Annual. Grow from seed planted in fall or very early spring. Be sure seed is ethically sourced as seed collection of annual plants can impact parent populations. Plants will readily self-sow and seeds are dispersed explosively when ripe. Use as an ephemeral groundcover or plant in the vegetable garden!

Miner's lettuce is a choice edible annual with small white flowers growing from the middle of fleshy, bright green, disclike leaves above a basal clump of spoon-shaped leaves. Reaching 1 ft. tall in ideal conditions, the entire plant makes a delicious salad green. When harvesting, be sure to leave enough plants to reseed the crop and feed the birds that relish this ephemeral beauty. There are many species of *Claytonia* native to the Pacific Northwest. Candyflower (*C. sibirica*) can create an edible groundcover in moist forest gardens west of the Cascades.

Miner's lettuce creates an edible emerald carpet of growth in the spring.

Clinopodium douglasii • Lamiaceae

Yerba buena

HABITAT/RANGE Open conifer forests and oak woodlands. Southern British Columbia to California and east to Montana. Grows mainly west of the Cascade Mountains, farther east through the Columbia River Gorge, but also occurs from south central British Columbia through northern Idaho, eastern Washington, and northeastern Oregon. Low to mid elevations.

SEASONAL INTEREST White flowers spring to summer. Evergreen in mild winters.

WILDLIFE VALUE Attracts bees, bumble bees, and other pollinators. Deer resistant.

CULTIVATION Part shade and well-drained, moist to seasonally dry soils. Water to establish. Drought tolerant but will need supplemental water in arid areas. A lovely groundcover for open woodland gardens. Can be used as a trailing plant along the edge of planters and shaded rock walls. Rooted stem sections can be separated from the plant and moved to new areas of the garden. Heavy mulch can choke out this low-growing plant.

Yerba buena is an attractive perennial groundcover that remains evergreen where winters are mild. The trailing stems bear ovate leaves with small white flowers at the leaf axils that attract bumble bees. This plant creeps nicely and unaggressively around shrubs, rooting at nodes along the stems and often forming mats. Aromatic when crushed, this is a well-known tea herb although it is important to get expert advice before consuming this or any wild plant.

Yerba buena is a lovely aromatic groundcover.

Coreopsis tinctoria • Asteraceae

Columbia coreopsis

HABITAT/RANGE Streambanks and riparian areas. Grows along major river systems in Oregon and Washington. Widely distributed in North America.

SEASONAL INTEREST Yellow flowers early summer to fall.

WILDLIFE VALUE Late-season nectar plant for pollinators. Attracts butterflies, bees, and other pollinators. Many solitary bees specialize on plants in the family Asteraceae. Seeds eaten by birds.

CULTIVATION Full sun and moist to wet soil. Does well in sandy or gravelly soil. Grows along rivers and other areas where it has "wet feet." Annual or biennial. Collect and sow seed from garden plants in fall or let plants reseed themselves. Deadheading spent flowers, as well as maintaining soil moisture in summer, prolongs bloom time. Mulch to retain soil moisture.

Columbia coreopsis is a colorful and long-blooming wildflower that is easily cultivated in sunny, moist sites. This plant can grow 3 ft. tall or more with branching stems and showy composite flowers. The ray florets, which look like the petals of the flower, are orange-yellow and their bases are often marked with a reddish blotch. The disc florets, which make up the center of the flower, are dark or reddish brown. This is a lovely plant for pollinator gardens where it provides nectar late in the growing season.

Columbia coreopsis is a colorful wildflower that grows along rivers and streams.

Cornus unalaschkensis • Cornaceae

Western bunchberry

HABITAT/RANGE Moist forests and woodlands. Alaska to Northern California and east to Montana at low to high elevations. Grows on both sides of the Cascade Mountains. Rare in California.

SEASONAL INTEREST White blooms in summer. Red fruits late summer to fall. Fall color, sometimes evergreen.

WILDLIFE VALUE Attracts bees and other pollinators. Fruits eaten by mammals and birds including gamebirds such as grouse. Provides cover for ground-nesting birds. Browsed by deer and elk.

CULTIVATION Full to part shade and moist, humus-rich, acidic soil. Water to establish. Some supplemental water may be needed after establishment in drier sites. Slow to establish. Spreads by rhizomes. Can grow under conifers. Use as a ground-cover in moist woodland gardens. Likes to grow on or near rotting wood; good for stumperies. Mulch.

Western bunchberry is a perennial groundcover that has it all: attractive foliage, showy flowers, and colorful fruits. The whorled leaves sometimes remain evergreen or add to fall colors by turning beautiful shades of red and yellow when cold weather arrives. The individual flowers are small and grow in tight clusters surrounded by large white bracts that look like petals. They attract pollinating insects and explosively release pollen on their visitors. The bright red, berrylike fruits are showy and a favored food of wildlife. This forest understory plant is a ground-hugging relative of dogwood trees and shrubs. Bunchberry (*Cornus canadensis*) is nearly identical and grows in northern portions of the region.

Western bunchberry is a gorgeous groundcover for moist woodland gardens that is showy both in flower and fruit.

Corydalis scouleri • Papaveraceae

Scouler's corydalis

HABITAT/RANGE Moist forests and shaded streambanks from southwestern British Columbia to west central Oregon. Occurs west of the Cascade Mountains from the coast to mountain foothills. Designated as threatened in Canada.

SEASONAL INTEREST Pink flowers spring to early summer.

WILDLIFE VALUE Attracts bees, butterflies, hummingbirds, and other pollinators. Seeds eaten by birds. Provides cover. Deer resistant.

CULTIVATION Full to part shade and moist to wet, humus-rich soil. Water to establish. Supplemental water may be needed after establishment if not in a suitably moist site. Tall and rhizomatous; give this species room to grow. Plants go dormant after setting seed. Use in wet, shady sites, woodland gardens, and riparian areas west of the Cascades. Mulch.

Corydalis scouleri is a lush herbaceous perennial perfectly suited for wet woodlands and shady coastal gardens. It grows 3 to 4 ft. tall with ferny foliage and showy racemes of spurred flowers that look like pink cornucopias. This species is a lovely plant for gardeners west of the Cascades. Gardeners east of the Cascades and in northern portions of the region can try golden corydalis (*C. aurea*), a wide-ranging and low-growing annual or biennial with yellow flowers that grows in both moist and dry soils. The genus *Corydalis* contains some rare and endemic species in the Pacific Northwest; be sure plants are ethically sourced.

Corydalis scouleri is a tall and beautiful shade-loving plant.

Darmera peltata • Saxifragaceae

Umbrella plant

HABITAT/RANGE Moist sites, grows along streams in the Coast Range from western Oregon to northwestern California and in the Sierra Nevada Mountains. Low to mid elevations. Grows west of the Cascade Mountains.

SEASONAL INTEREST Pink to white flowers in spring to early summer depending on location. Large umbrellalike leaves. Fall color.

WILDLIFE VALUE Attracts pollinators. Provides cover. Helps shade and stabilize streambanks.

CULTIVATION Part shade to sun and moist to wet soil. Water to establish and continue to provide supplemental water. Prefers cool summers and partly shaded sites with consistent moisture. Does not do well in hot, dry areas. A large plant; give it plenty of space. Plant along streams and pond edges, in boggy places, and moist gardens. Can be grown in containers. Mulch.

Darmera peltata is a unique and distinctive plant that makes for a beautiful ornamental in moist gardens. In spring, leafless stems up to 5 ft. tall, though usually shorter, bear showy clusters of pink to white flowers. Later, umbrellalike leaves emerge and continue to expand reaching over 1 ft. across and forming 3 ft. tall clumps. In sites with consistent moisture these leaves usually persist through summer, turning shades of yellow and red in fall before dying back to the thick rhizomes. This tropical-looking herbaceous perennial is the only species in its genus and has a limited range confined to Oregon and California; make sure plants are ethically sourced.

Darmera peltata has tall clusters of flowers that bloom before large umbrellalike leaves fully unfurl.

Delphinium nuttallianum • Ranunculaceae

Upland larkspur

HABITAT/RANGE Dry, rocky sites, meadows, grasslands, shrub-steppe, eastside forests, and mountain valleys and slopes. British Columbia to California and east to the Rocky Mountains at low to high elevations. Grows mainly east of the Cascade Mountains.

SEASONAL INTEREST Dark to pale purplish blue flowers in spring to summer relative to elevation.

WILDLIFE VALUE Attracts hummingbirds, bees, butterflies, and other pollinators. Moth host plant. Toxic.

CULTIVATION Sun to light shade and seasonally dry, well-drained soils. Water only to establish. Drought tolerant. Do not plant in areas with continual irrigation. Goes dormant after setting seed. Slugs are very attracted to this plant. Great for rock gardens. This is a highly variable species; favor planting ethically propagated starts from locally sourced seed.

Upland larkspur is a showy species of *Delphinium* for gardeners in open pine forests, shrub-steppe, and sunny, rocky areas east of the Cascades. Adapted to drier climates, this attractive species can be shorter and less robust than other delphiniums, growing 6 to 15 in. tall with mostly basal leaves. The flowers have purplish blue sepals and nectar-bearing spur, with white, purple-lined or purple-tinged petals. Plant more than one of this drought-tolerant herbaceous perennial as upland larkspur looks loveliest blooming in groups. There are many species of *Delphinium* native to the region. Menzies' larkspur (*D. menziesii*) is similar and a better choice for gardeners west of the Cascades. Red larkspur (*D. nudicaule*), a hummingbird favorite with scarlet flowers, grows from southwestern Oregon to central California.

Upland larkspur is a widespread, highly variable species and a colorful drought-tolerant perennial.

Delphinium trolliifolium • Ranunculaceae

Columbia larkspur

HABITAT/RANGE Moist woodlands and shaded streambanks west of the Cascade Mountains and in the Columbia River Gorge at low elevations from southwestern Washington to Northern California.

SEASONAL INTEREST Blue flowers spring to early summer.

WILDLIFE VALUE Attracts hummingbirds, butterflies, bees, bumble bees, and other pollinators. Moth host plant. Toxic.

CULTIVATION Full to part shade and moist, humus-rich soil. Water to establish and continue to supply supplemental water as needed. Goes dormant after setting seed. Protect from slugs. Use in shaded woodland gardens, riparian areas, rain gardens, and pollinator gardens. Mulch.

Delphinium trolliifolium is a shade-loving herbaceous perennial that grows 2 to 5 ft. tall with bright blue flowers and deeply lobed leaves. This plant is prized by hummingbirds and bumble bees, making it a great addition to pollinator gardens. It also makes a beautiful cut flower, although the hummingbirds might protest. Columbia larkspur is a knockout in the garden but has a fairly limited range, occurring primarily in western Oregon. Luckily for gardeners, there are several species of delphiniums native to the region that are amply gorgeous in cultivation; plant the ones found near you. Pale larkspur (*D. glaucum*) is another tall and robust species that grows in moist sites at mid to high elevations throughout the region.

Columbia larkspur is a colorful shade-loving perennial that attracts hummingbirds and bumble bees.

Dicentra formosa • Papaveraceae

Pacific bleedingheart

HABITAT/RANGE Moist forests and streambanks. Southern British Columbia to California with disjunct populations in southeastern Washington, northeastern Oregon, and Idaho. Low to mid elevations. Grows mainly west of the crest of the Cascade Mountains.

SEASONAL INTEREST Purplish pink flowers spring to summer, occasionally again in fall. Seasonal groundcover.

WILDLIFE VALUE Attracts bees, butterflies, hummingbirds, and other pollinators. Host plant for the Clodius parnassian butterfly. Attracts beneficial insects. Seeds have a fatty protein package, called an elaiosome, that attracts ants who disperse the seed. Provides cover for small animals. Deer resistant.

CULTIVATION Full to part shade and moist, humus-rich soil. Water to establish. Tolerates seasonally dry conditions if properly sited in shady sites west of the Cascades. Dies back in hot summer weather, sometimes reemerging to bloom again in fall. Supplemental water may keep foliage green through the summer and prolong bloom time. Spreads by seed and brittle rhizomes, aggressively in some sites. Protect from slugs. Mulch.

Pacific bleedingheart is a widely cultivated herbaceous perennial known for its nodding, pink, heart-shaped flowers and fernlike foliage. It is a low-growing plant, 6 to 18 in. tall, that creates an attractive seasonal groundcover in moist, shady woodlands and looks lovely interspersed with ferns and other understory plants. *Dicentra formosa* is a parent plant for many hybrids and cultivars commonly sold in nurseries. Choose plants wisely and always favor planting true natives.

Pacific bleedingheart is an easily recognized, unique wildflower and a popular plant for moist woodland gardens.

Dichelostemma congestum • Asparagaceae

Ookow

HABITAT/RANGE Meadows, grasslands, oak woodlands, and rocky prairies. Puget Sound in Washington south to California and east through the Columbia River Gorge. Low to mid elevations. Grows mainly west of the Cascade Mountains.

SEASONAL INTEREST Purple flowers late spring to early summer.

WILDLIFE VALUE Attracts hummingbirds, bees, and other pollinators. Favored nectar plant of butterflies, particularly swallowtail butterflies. Bulbs eaten by mammals. Browsed by deer.

CULTIVATION Sun to light shade and well-drained soil. Prefers vernally moist soils that are dry in summer. Water only to establish. Drought tolerant. Do not plant in regularly irrigated areas. Goes dormant after setting seed. Increases slowly by corm offsets and seed. Takes a few years from seed to flower. Mulch lightly.

Ookow's bluish purple flowers signal the beginning of summer. They bloom in tight clusters atop leafless stems that grow 1 to 3 ft. tall from bulblike corms with a few linear, grasslike, basal leaves. This perennial bulb creates a lovely color accent in sunny xeric gardens and its sturdy but slender stems grow nicely through other perennials and grasses. Bluedicks (*Dichelostemma capitatum*) is similar and ranges from southern Oregon through California. Both plants are excellent choices for butterfly and pollinator gardens. As with any native bulb, be sure initial starts are seed propagated and not taken from the wild.

Ookow is a tall and colorful wildflower loved by swallowtail butterflies.

Dodecatheon spp./*Primula* spp. • Primulaceae

Shooting star

HABITAT/RANGE There are many species found throughout the region that are adapted to various habitats including coastal areas, mountain meadows, streambanks, wetlands, open forests, oak woodlands, and rocky slopes at low to high elevations.

SEASONAL INTEREST Pink or white flowers in spring or summer.

WILDLIFE VALUE Attracts bumble bees and other pollinators. Buzz-pollinated. Attracts beneficial insects. Butterfly host plant.

CULTIVATION Most shooting stars grow in sunny to lightly shaded, moist to vernally moist sites. Plants go dormant after setting seed. Some species prefer dry summers and will need soils to dry out after they go dormant. Plant species found in your area and habitat. Make sure plants are ethically propagated.

Shooting stars are interesting and colorful perennials. Their flowers have reflexed petals and bloom on leafless stalks over a basal rosette of leaves. There are many native species that can be lovely in gardens. Tall mountain shooting star (*Dodecatheon jeffreyi/Primula jeffreyi*) grows in moist places with purple, sometimes white, flowers. White shooting star (*D. dentatum/P. latiloba*) spreads by rhizomes, creating lovely clumps. Broad-leaved shooting star (*D. hendersonii/P. hendersonii*) is a good species for seasonally dry sites west of the Cascades. Pretty shooting star (*D. pulchellum/P. pauciflora*) is widespread with multiple variations, one coastal. Cultivars of *D. meadia/P. meadia*, a native of eastern North America, are commonly available; choose plants wisely and plant species native to the Pacific Northwest. This genus is classified as either *Dodecatheon* or *Primula* depending on the source.

Poet's shooting star (*Dodecatheon poeticum/Primula poetica*) in the sunset light.

Drymocallis glandulosa • Rosaceae

Sticky cinquefoil

HABITAT/RANGE Forest openings, slopes, and meadows. British Columbia to California and east to Montana at low to high elevations. Widespread and highly variable with multiple subspecies.

SEASONAL INTEREST Bright to pale yellow or white flowers in spring to midsummer relative to elevation.

WILDLIFE VALUE Important for native bees, especially mason and carpenter bees, as well as a few species of solitary specialist bees. Attracts pollinating and beneficial insects. Butterfly host plant. Browsed by deer, elk, and small mammals.

CULTIVATION Sun to part shade and moist to seasonally dry, well-drained soil. Grows in a variety of soil types. Water to establish. Drought tolerant once established if properly sited. This is a variable species; use starts grown from locally sourced seed. An excellent plant for pollinator gardens and wildflower meadows.

The true beauty of sticky cinquefoil lies in its importance to native bees. It is often overlooked as a garden plant, yet it can be quite charming. This wildflower grows up to 2.5 ft. tall and favors partly sunny spots with good drainage, decent soils, and seasonal moisture. Sticky cinquefoil produces multiple yellow to white flowers on branching stems and attractive compound leaves with toothed leaflets that resemble strawberry leaves. As its name implies, it is glandular and sticky to the touch. This herbaceous perennial was previously classified in the genus *Potentilla*, in which there are many species suitable for gardens with similar qualities and benefits to native pollinators.

Sticky cinquefoil is a cheerful, drought-tolerant wildflower loved by native bees.

Epilobium canum • Onagraceae

Hummingbird trumpet

HABITAT/RANGE Rocky slopes and ridges from southwestern Oregon through California to the southwestern United States. Grows along seasonal creeks and seeps in hotter and drier portions of its range. Low to high elevations.

SEASONAL INTEREST Red flowers summer into fall. Seasonal groundcover.

WILDLIFE VALUE Attracts hummingbirds and helps fuel their migration south. Late-season nectar source for bees, butterflies, and other pollinators. Moth host plant.

CULTIVATION Sun to part shade and well-drained soils. Water to establish. Drought tolerant once established. Benefits from occasional supplemental water in hot, sunny sites. Do not overwater. Rhizomatous and fairly low growing; use as a groundcover. New growth often does not emerge until late spring. Great for rock gardens, parking strips, pollinator plantings, and hummingbird gardens. Grows well in containers. Responds well to gravel mulch.

Hummingbird trumpet is an attractive herbaceous, though somewhat woody, perennial. It ranges farther into the Southwest than it does the Pacific Northwest, coming up through California into a small portion of southwestern Oregon. It is widely available and a popular garden plant in the region that blooms continuously summer into fall when most plants are already dormant, making it an important nectar source for pollinators. It grows 4 to 19 in. tall and produces showy, bright red to orange-red, trumpet-shaped flowers that contrast its gray-green foliage. True to its name, it is a magnet for hummingbirds. Also called California fuchsia, it was previously classified as *Zauschneria californica*.

Hummingbird trumpet provides vibrant color and a nectar source for pollinators late in the growing season.

Erigeron glacialis • Asteraceae

Glacier fleabane

HABITAT/RANGE Moist meadows, open woodlands, and streambanks in mountainous areas. Alaska to California and east to the Rocky Mountains at mid to high elevations. Grows on both sides of the Cascade Mountains.

SEASONAL INTEREST Dark to pale purple flowers in spring to late summer relative to elevation.

WILDLIFE VALUE Attracts butterflies, bees, and other pollinators. Attracts beneficial insects. Butterfly host plant. Specialist bee host. Birds often eat seeds of plants in the family Asteraceae.

CULTIVATION Sun to part shade and moist, humus-rich soil. Water to establish and continue to provide supplemental water as needed. Deadhead to prolong bloom time. Use in butterfly gardens. Can be grown in containers. Mulch.

Glacier fleabane is an attractive perennial that grows well at any elevation despite the fact it naturally tends to occur at higher altitudes. It has composite flowers consisting of purple ray flowers surrounding a dense cluster of tiny yellow disc flowers that create a perfect platform for butterflies to land on. Growing 1 to 2 ft. tall in moist sites, garden plants make an excellent cut flower. This plant was formerly classified as a variation of *Erigeron peregrinus*. A number of species of *Erigeron* native to the Pacific Northwest make lovely additions to pollinator gardens. Seaside daisy (*E. glaucus*) is a short and showy coastal denizen with purple flowers. Showy fleabane (*E. speciosus*) has blue-violet flowers and is easily grown on either side of the Cascades.

Glacier fleabane is a very hardy, large-flowered daisy.

Erigeron linearis • Asteraceae

Desert yellow daisy

HABITAT/RANGE Dry, rocky plains and foothills from southern British Columbia to California and east to Montana. Low to mid elevations. Grows east of the Cascade Mountains.

SEASONAL INTEREST Yellow flowers in spring to midsummer depending on location.

WILDLIFE VALUE Attracts butterflies, bees, and other pollinators. Attracts beneficial insects. Butterfly host plant. Specialist bee host. Seeds may be eaten by birds.

CULTIVATION Full sun and rocky, dry, well-drained soil. Water to establish. Drought tolerant once established. Do not plant in irrigated areas. Needs sharp drainage. Use in rock gardens and sunny pollinator plantings. Not easily grown west of the Cascades.

Erigeron linearis is a truly beautiful plant for dry, sunny gardens. This drought-tolerant perennial puts on quite a show when in bloom, producing masses of yellow flowers over a tidy cushion of blue-gray, linear leaves. It is short in stature, growing 1 ft. tall or less from a woody base, and makes a compact and colorful rock garden plant. Thread-leaf fleabane (*E. filifolius*) also grows in dry places east of the Cascades and has similar foliage but is less compact, with flowers that range from white to pink or blue in color. These sun-loving plants look lovely growing with buckwheats, desert parsleys, bunchgrasses, and rabbitbrush.

Erigeron linearis has attractive foliage and bright yellow flowers.

Eriogonum compositum • Polygonaceae

Arrowleaf buckwheat

HABITAT/RANGE Rocky slopes, meadows, shrub-steppe, and open forests. Washington to California and east to Idaho at low to high elevations. Grows mainly east of the Cascade Mountains with a few populations in southwestern Oregon, the Willamette Valley, and throughout the Columbia River Gorge.

SEASONAL INTEREST Creamy white or yellow flowers in spring to midsummer relative to elevation.

WILDLIFE VALUE Important for beneficial insects. Flowers prized by pollinators. Host plant for multiple species of butterflies. Seeds eaten by birds. Browsed by deer and small animals.

CULTIVATION Sun to light shade and rocky, well-drained soil. Tolerates poor soils with good drainage. Water to establish. Drought tolerant once established. Do not plant in regularly irrigated gardens. Can be grown west of the Cascades.

Native buckwheats compose a genus of many variable species that are not only beautiful but also very important for birds, pollinators, butterflies, and beneficial insects. Arrowleaf buckwheat is a woody perennial that boasts large, showy, cloudlike clusters of yellow to creamy white flowers, which turn into colorful seed heads on stout, leafless stems 1 to 2 ft. tall. Usually mat-forming with two-toned basal leaves that are green above and woolly white below, this is an attractive plant that does well in cultivation given adequate sun and well-drained soils. Great for insectary plantings on organic farms and vineyards east of the Cascades.

Arrowleaf buckwheat is very drought tolerant and showy in bloom.

Eriogonum elatum • Polygonaceae

Tall buckwheat

HABITAT/RANGE Rocky and sandy slopes, grasslands, shrub-steppe, mountain ridges, and open forests. Northern Washington to California and east to Idaho. Grows at low to mid elevations east of the Cascade Mountains with populations also in the Klamath Mountains.

SEASONAL INTEREST White flowers, that are pink in bud, summer to early fall.

WILDLIFE VALUE Very important for beneficial insects and pollinators. Host plant for multiple species of butterflies. Provides food for birds.

CULTIVATION Sun to light shade and dry, rocky, well-drained soils. Tolerates poor soils. Water to establish. Drought tolerant once established. Do not plant in places with continual irrigation. Use in rock gardens, xeric gardens, and sunny pollinator and insectary plantings east of the Cascades.

Tall buckwheat is essential to pollinating insects in hot, dry areas in late summer. Its ability to bloom in dry soils during desiccating heat is truly remarkable. It has proven to be one of the best plants for attracting beneficial insects and should be included in organic farm and vineyard insectary plantings. With wiry, leafless stems growing to around 3 ft. tall from a basal clump of large ovate leaves, the stature and branching structure of this drought-tolerant perennial make it appealing and worthy of a place in the garden. The flowers add late-summer color and are pink in bud, white in bloom, and fade to a rusty red. Naked buckwheat (*Eriogonum nudum*) is similar but with larger flower clusters and is a better choice for gardeners west of the Cascades.

Tall buckwheat is an important plant for pollinators and beneficial insects.

Eriogonum heracleoides • Polygonaceae

Parsnip-flowered buckwheat

HABITAT/RANGE Open, rocky sites, shrub-steppe, pine forests, and mountain ridges. British Columbia to northeastern California and east to the Rocky Mountains at low to high elevations. Grows east of the Cascade Mountains.

SEASONAL INTEREST Cream-colored flowers, often tinged rosy red in bud, late spring to summer.

WILDLIFE VALUE Important for beneficial insects. Flowers prized by pollinators. Host plant for multiple species of butterflies. Seeds eaten by birds and small animals. Browsed by deer and elk.

CULTIVATION Sun to light shade and well-drained soil. Likes rocky to loamy soils but tolerates poor soils with good drainage. Water to establish. Drought tolerant once established. Do not plant in gardens with continual irrigation. A beautiful plant for sunny xeric gardens. Grows well in containers. Mulch lightly.

Parsnip-flowered buckwheat is an attractive shrubby perennial that thrives in hot, dry, sunny sites. It has compound umbels of showy cream-colored flowers that are often colorfully tinged with rose-red hues when in bud. This plant can bloom for a long time and grows 1 to 2 ft. tall with linear leaves and leaflike bracts arranged in whorls and loose rosettes on woody stems. A great butterfly plant perfect for insectary, pollinator, and restoration plantings east of the Cascades.

The cream-colored flowers of *Eriogonum heracleoides* may display attractive shades of rosy red when budded.

Eriogonum strictum • Polygonaceae

Strict buckwheat

HABITAT/RANGE Sandy and rocky sites, shrub-steppe, and pine forests. Northern Washington to Northern California and east to Montana at low to mid elevations. Grows east of the Cascade Mountains.

SEASONAL INTEREST White, pink, or yellow flowers summer to early fall.

WILDLIFE VALUE Important for beneficial insects. Late-season pollen and nectar source for pollinators. Host plant for multiple species of butterflies. All species of *Eriogonum* are important butterfly plants.

CULTIVATION Sun to light shade and dry, rocky, well-drained soil. Tolerates poor soils with good drainage. Water to establish. Drought tolerant once established. Do not plant in places with continual irrigation. Use in rock gardens, xeric gardens, parking strips, and sunny butterfly gardens.

Strict buckwheat is noteworthy in its ability to bloom profusely late in the growing season in dry, sunny sites, providing vital food for pollinators, as well as color in xeric gardens. It is a woody perennial with woolly foliage and a distinctly silvery gray appearance. Flowers can be white, pinkish, or yellow, depending on the variation, and bloom in clusters on branching stalks that grow to around 1 ft. tall. Snow buckwheat (*Eriogonum niveum*) is nearly identical but blooms a little later and is a bit taller. Both of these drought-tolerant species attract beneficial insects and should be included in organic farm and vineyard insectary and pollinator plantings east of the Cascades.

Strict buckwheat is a showy, late-blooming perennial that provides forage for pollinators in sunny, dry areas.

Eriogonum umbellatum • Polygonaceae

Sulfur buckwheat

HABITAT/RANGE This species has many variations adapted to a range of habitats from sagebrush deserts to alpine ridges. Southern British Columbia through California and east to the Rocky Mountains at low to high elevations. Grows on both sides of the Cascade Mountains, though primarily in and east of the mountains.

SEASONAL INTEREST Yellow to cream flowers, tinged orange to red in bud, late spring to early fall depending on location.

WILDLIFE VALUE Important for beneficial insects and pollinators. Butterfly larval host. Seeds eaten by birds and small animals. Browsed by grouse, quail, elk, and deer.

CULTIVATION Sun to light shade and well-drained soils. Tolerates poor soils with good drainage. Water to establish. Drought tolerant once established. Do not plant in places with continual irrigation. This is a variable species adapted to differing conditions; plant starts grown from locally sourced seed. Great for rock gardens, butterfly gardens, and pollinator plantings. Mulch can smother this low-growing plant.

Sulfur buckwheat is a beautiful woody perennial with showy clusters of flowers that are bright yellow to cream, with tints of vibrant orange to red both in bud and as flowers fade. It is low growing and usually mat forming, reaching about 2 ft. across and 1 ft. tall depending on the variation, with basal leaves that may remain evergreen or turn red as they fade giving the plant some fall color. Scabland buckwheat (*Eriogonum sphaerocephalum* var. *sublineare*), Douglas's buckwheat (*E. douglasii*), and thyme-leaved buckwheat (*E. thymoides*) are also attractive, tidy, mat forming, and drought tolerant.

Sulfur buckwheat is a lovely plant for sunny xeric gardens and pollinator plantings.

Eriophyllum lanatum • Asteraceae

Oregon sunshine

HABITAT/RANGE Dry, open, rocky places. British Columbia to California and east to the Rocky Mountains at low to high elevations. Widespread in the region. Grows on both sides of the Cascade Mountains.

SEASONAL INTEREST Yellow flowers in spring to late summer relative to elevation.

WILDLIFE VALUE Loved by butterflies, bees, and other pollinators. Attracts beneficial insects. Host plant for the painted lady butterfly. Nectar source for the endangered Fender's blue butterfly. Many solitary bees specialize on plants in the family Asteraceae. Seeds eaten by birds. Deer resistant.

CULTIVATION Sun to light shade and well-drained, seasonally dry soil. Tolerates poor soils but needs sharp drainage. Water only to establish. Drought tolerant once established. Do not plant in sites with continual irrigation. Spreads by seed. Great for rock gardens, butterfly gardens, parking strips, and xeric gardens. Can be used in planters. No mulch or a light gravel mulch.

Butterflies will be delighted if you plant this colorful and attractive herbaceous perennial and so will you. The only way to go wrong with Oregon sunshine is to overwater it. This plant needs well-drained soils that are moist in spring and dry in summer along with a sunny site to fit its disposition. It grows 4 to 24 in. tall with profusions of yellow, daisylike flowers. There are many intergrading variations of this species in the region with foliage ranging from leathery green to white and woolly, and plant habits varying from tight and compact to loose and sprawling.

Oregon sunshine is a bright and beautiful plant for dry, sunny spots in the butterfly garden.

Erythranthe guttata • Phrymaceae

Yellow monkeyflower

HABITAT/RANGE Wet places, seeps, vernal pools, riparian areas, coastal wetlands, and ditches. Alaska south through California and east beyond the Rocky Mountains at low to high elevations. Widespread. Grows on both sides of the Cascade Mountains.

SEASONAL INTEREST Yellow flowers in spring to early fall relative to elevation and soil moisture.

WILDLIFE VALUE Attracts bees, butterflies, hummingbirds, and other pollinators. Butterfly host plant. Attracts beneficial insects. Specialist bee host. Browsed by deer.

CULTIVATION Sun to part shade and moist to wet soil. Annual or perennial. Let plants reseed. Spreads by seed or rhizomes. Deadhead or cut back to produce more blooms. Maintaining soil moisture prolongs bloom time and increases growth. Dies back if soils dry out. Use in riparian areas, rain gardens, wet meadows, and irrigated butterfly gardens.

Yellow monkeyflower is a fast-growing wildflower that colors the edges of waterways and wet areas with bright splashes of yellow. It is a variable species and may grow either as an annual or a rhizomatous herbaceous perennial. Its habit can be trailing or upright, growing anywhere from 3 in. to 3 ft. high with succulent stems and long-blooming, large, yellow tubular flowers. Formerly classified as *Mimulus guttatus*, this is one of the showiest and most common species of monkeyflower in the region. Take care if collecting wild seed as there are rare species that may look similar and collecting seeds of annual plants can have a big impact on wild parent populations. Plants will readily reseed themselves in moist gardens.

Yellow monkeyflower adds bright splashes of color to wet places.

Erythranthe lewisii • Phrymaceae

Great purple monkeyflower

HABITAT/RANGE Wet places, seeps, and streambanks in mountainous areas from Alaska to California and east to the Rocky Mountains. Mid to high elevations.

SEASONAL INTEREST Purplish pink flowers in summer.

WILDLIFE VALUE Attracts hummingbirds, bees, butterflies, and other pollinators. Butterfly host plant. Attracts beneficial insects. Specialist bee host.

CULTIVATION Sun to part shade and cool, moist to wet, well-drained soil. A showy species worth cultivating though it prefers growing along mountain streams and may be shorter lived and less robust at lower elevations. Will not do well in hot, dry areas. Rhizomatous but maintains a clumping habit. Deadhead to encourage more blooms. Mulch lightly.

Great purple monkeyflower is a large-flowered, brightly colored, and very attractive herbaceous perennial with a clumping, erect habit that can grow up to 3 ft. tall, though usually it is shorter. In summer the plant produces deep purplish pink tubular flowers with yellow throats that attract hummingbirds. Formerly classified as *Mimulus lewisii*, this is undoubtedly one of the showiest monkeyflowers in the region. Also loved by hummingbirds is scarlet monkeyflower (*Erythranthe cardinalis*), a perennial with bright red to orange-red flowers that grows at lower elevations from west central Oregon to California.

Great purple monkeyflower puts on a showy display along mountain streams with its large purplish pink flowers.

Erythronium grandiflorum • Liliaceae

Glacier lily

HABITAT/RANGE Oak woodlands, open forests, and meadows. British Columbia to Northern California and east to the Rocky Mountains at low to high elevations. Grows on both sides of the Cascade Mountains.

SEASONAL INTEREST Yellow flowers in early spring to summer relative to elevation.

WILDLIFE VALUE Attracts bees, bumble bees, butterflies, hummingbirds, and other pollinators. Browsed by deer, elk, and small mammals. A favored food of bears.

CULTIVATION Part shade to sun and well-drained, vernally moist, humus-rich soil. Water to establish. Drought tolerant once established if properly sited. Prefers soils that are moist in spring but dry in summer. Goes dormant after setting seed. Collect and sow seed from garden plants to increase your population as this species looks gorgeous growing in multitudes. Takes a few years from seed to flower. Plant under oak trees, along forest edges, and in rock gardens. Grows best east of the Cascades or in higher elevation gardens. Mulch lightly.

Glacier lily is a cheerful perennial bulb that has large, nodding, yellow flowers with flared tepals and anthers of unusual size. Short in stature, it grows from 4 to 12 in. tall with one to three flowers on leafless stalks and a pair of large, uniformly green, basal leaves. This plant is loved by bumble bees and other native bees looking for an early-season food source. There are a few variations of this species in the region and gardeners will be most successful planting starts propagated from locally and ethically sourced seed. Never dig these beauties from the wild.

Glacier lily's flared yellow flowers are ephemeral eye candy.

Erythronium oregonum • Liliaceae

Giant fawn lily

HABITAT/RANGE Oak woodlands, open forests, prairies, and rocky balds. Southwestern British Columbia through the Puget Trough to western Oregon. Rare in California. Grows mainly at low elevations west of the Cascade Mountains.

SEASONAL INTEREST White flowers in spring.

WILDLIFE VALUE Attracts bees, bumble bees, butterflies, humming-birds, and other pollinators. Browsed by deer and small mammals.

CULTIVATION Part shade to sun and moist, well-drained, humus-rich soil. Water to establish. Tolerates seasonally dry conditions once established if properly sited. Goes dormant after setting seed. Allow plants to self-sow to increase garden population. Use in woodland gardens or rock gardens with adequate moisture and some shade. Best grown in gardens west of the Cascades. Protect from slugs. Mulch lightly.

Giant fawn lily is a lovely perennial bulb that has one to three large, white, nodding flowers with flared tepals on leafless stalks about 1 ft. tall. Not only are the flowers attractive, but so are the large basal leaves that are mottled with green and rusty brown hues. Coast fawn lily (*Erythronium revolutum*) also makes a nice addition to gardens west of the Cascades. It is a pink-flowered species with strongly mottled leaves that grows along the coast and is considered sensitive in parts of its range. As with all native bulbs, make sure garden starts are ethically sourced and seed propagated, never dug from the wild. Rare, endemic, and threatened species of *Erythronium* exist in the region.

Giant fawn lily is a lovely spring bloomer for gardens west of the Cascades.

Eschscholzia californica • Papaveraceae

California poppy

HABITAT/RANGE Roadsides and fields. Occurs from southern British Columbia to Mexico and east to Idaho, typically at low elevations. Grows mainly west of the Cascade Mountains and east through the Columbia River Gorge. Considered exotic north of the Columbia River Gorge and introduced in parts of the region. California state flower.

SEASONAL INTEREST Bright orange flowers late spring to early fall.

WILDLIFE VALUE Attracts bees, bumble bees and other pollinators. Specialist bee host. Butterfly host plant. Seeds eaten by birds and small animals. Deer resistant.

CULTIVATION Sun to light shade and well-drained soil. Water to establish. Drought tolerant once established. May go dormant by late summer in hot, dry sites. Occasional summer water helps prolong bloom time and keep plants green. Deadheading also extends bloom time. Annual or perennial. Easy to grow from seed. Explosive seed capsules will spread it around your garden. Plant in sunny xeric gardens. Does well in planters.

California poppy is an attractive garden plant that is long blooming, drought tolerant, and loved by bumble bees. It grows to about 20 in. tall with bluish gray, fernlike foliage and large, showy, bright orange to yellow flowers and is a common component of wildflower seed mixes sold throughout the United States. It has escaped cultivation in many places and there is some discrepancy about its native status in parts of the Pacific Northwest. Cultivars of this easily grown wildflower come in a range of colors. A stunning plant for sunny, dry pollinator gardens and parking strips.

California poppy is a brightly colored, drought tolerant garden favorite.

Fragaria vesca • Rosaceae

Woodland strawberry

HABITAT/RANGE Woodlands, open forests, streambanks, and meadows. British Columbia to California and east across North America at low to fairly high elevations. Grows on both sides of the Cascade Mountains.

SEASONAL INTEREST White flowers in spring to early summer. Edible red fruits in summer.

WILDLIFE VALUE Attracts bees, butterflies, and other pollinators. Butterfly host plant. Attracts beneficial insects. Fruits eaten by birds. Browsed by deer, elk, and small mammals.

CULTIVATION Part shade to sun and moist, well-drained, humus-rich soil. Water to establish. Tolerates dry summer conditions once established but prefers some shade and supplemental water in hot, dry areas. Spreads quickly by runners that are easily controlled. Use as a groundcover under shrubs and along borders or rock walls where it can trail down, rooting and growing in between rocks. A good plant for green roofs. Mulch lightly.

One of our best groundcovers, woodland strawberry is easily grown, quick to establish, and tolerates a variety of conditions. This stoloniferous herbaceous perennial creeps beautifully through the garden, covering and holding soils. It has attractive rosettes of veined leaves and white flowers borne above the foliage, which develop into small but delicious edible fruits. Coastal strawberry (*Fragaria chiloensis*) grows along the coast, is generally evergreen, and is a popular garden plant with commonly available cultivars and hybrids. Mountain strawberry (*F. virginiana*) is similar to *F. vesca* though its flowers are generally found below the foliage. All these species make excellent edible groundcovers.

Woodland strawberry is an edible groundcover with showy white flowers and small, delicious fruits.

Fritillaria affinis • Liliaceae

Checker lily

HABITAT/RANGE Oak woodlands, open forests, prairies, and rocky slopes. Southern British Columbia to California, disjunct in eastern Washington and Idaho. Low to mid elevations. Grows mainly from the eastern foothills of the Cascade Mountains west to the coast.

SEASONAL INTEREST Green, purplish brown, and yellow checkered flowers in spring to early summer.

WILDLIFE VALUE Attracts pollinators. Bulbs eaten by mammals.

CULTIVATION Sun to part shade and seasonally moist, well-drained soil. Give this plant some shade in hot, dry areas. Prefers soils that are moist in spring and dry in summer. Do not plant in regularly irrigated sites. Water to establish. Goes dormant after setting seed. Takes a few years to flower from seed. Disturbing mature bulbs encourages small bulblets to separate and create new plants. Mulch lightly.

This beautiful and interesting perennial bulb has whorled leaves and a sturdy stem that usually grows 1 to 2 ft. tall bearing one to five nodding, checkered flowers. The large bell-shaped flowers are curiously colored and have a fetid smell designed to attract pollinating flies. The smell is not noticeable in the garden, but it's not recommended as a cut flower. Plant this lovely lily in rock gardens and under oak trees. Chocolate lily (*Fritillaria atropurpurea*) is similar and grows chiefly east of the Cascades from Oregon south. Native lilies like these are under pressure from development and invasive species and are becoming increasingly uncommon. Make sure plants are ethically propagated.

Checker lily boasts a unique flower mottled with green, purplish brown, and yellow.

Gaillardia aristata • Asteraceae

Blanketflower

HABITAT/RANGE Open places, meadows, grasslands, and shrub-steppe. Central British Columbia to central Oregon and east to Minnesota at low to mid elevations. Grows mainly east of the Cascade Mountains.

SEASONAL INTEREST Yellow flowers with reddish purple centers summer to fall.

WILDLIFE VALUE Attracts bees, butterflies, and other pollinators. Attracts beneficial insects. Many solitary bees specialize on plants in the family Asteraceae. Moth host plant. Seeds eaten by birds.

CULTIVATION Sun and well-drained, seasonally dry to moist soils. Can grow in a variety of soil types including rocky or sandy soils. Prone to root rot in poorly drained soils. Water to establish. Drought tolerant once established but benefits from occasional summer water. Deadheading encourages blooms and prolongs bloom time. Sometimes a short-lived perennial; let it reseed itself or save seed from garden plants to sow in the future. Easy from seed. Use in xeric gardens, rock gardens, parking strips, and sunny pollinator gardens.

Blanketflower is a popular garden plant loved for its showy, colorful flowers and drought tolerance. This herbaceous perennial grows around 2 ft. tall with composite flowers consisting of reddish purple disc flowers surrounded by yellow ray flowers that are sometimes stained red at the base. This pretty flower blooms late into the season and is popular with bees and butterflies, making it a great choice for pollinator plantings. Many well-known cultivars are derived from this beautiful plant; choose plants wisely and favor planting true natives.

Blanketflower is a colorful, long-blooming, and drought-tolerant perennial.

Geranium viscosissimum • Geraniaceae

Sticky purple geranium

HABITAT/RANGE Meadows, open slopes, and open forests. British Columbia to California and east through the Rocky Mountains, from foothills to fairly high elevations. Grows east of the Cascade Mountains.

SEASONAL INTEREST Deep pink to lavender flowers late spring to summer.

WILDLIFE VALUE Attracts bees, butterflies, and other pollinators. Seeds eaten by birds and small mammals. Browsed by deer and elk.

CULTIVATION Sun to part shade and moist to seasonally dry, well-drained soil. Water to establish. Tolerates seasonally dry conditions once established if properly sited. Supplemental water may prolong bloom time and keep plants from going dormant in the heat of summer. Mulch lightly.

Geranium viscosissimum is an herbaceous perennial best suited for gardens east of the Cascades. Growing between 1 to 3 ft. tall, it has deep pink to lavender flowers with dark nectar lines designed to direct pollinator landings, deeply lobed leaves, and is covered with sticky, glandular hairs. Like some other sticky plants, it is thought to be protocarnivorous, meaning it can absorb nutrients from insects caught on the surface of the plant. This ability may help it survive in nutrient-poor soils. For gardeners west of the Cascades, western geranium (*G. oreganum*) is similar and grows in meadows and woodlands in western Oregon.

Geranium viscosissimum is a lovely drought-tolerant plant.

Geum triflorum • Rosaceae

Prairie smoke

HABITAT/RANGE Meadows, rocky slopes, open forests, and shrub-steppe. Yukon to California and east to the Great Lakes at low to high elevations. Grows in the Puget Sound and Olympic Mountains, otherwise mainly east of the Cascade Mountains.

SEASONAL INTEREST Pink flowers in spring to summer depending on location. Showy seed heads.

WILDLIFE VALUE Attracts bees, butterflies, and other pollinators. Moth host plant. Seeds eaten by birds.

CULTIVATION Sun to light shade and moist to seasonally dry, well-drained soil. Prone to root rot in poorly drained soils. Water to establish. Drought tolerant once established. Supplemental water will be needed in hot, dry sites to keep foliage green through summer. Spreads slowly by rhizomes. Garden plants can be divided. Use in rock gardens, sunny garden borders, meadows, and green roofs. Plants shed leaves in a way that is self-mulching.

Prairie smoke is a low-growing, rhizomatous, drought-tolerant perennial with fuzzy fernlike foliage. The nodding flowers have reddish pink sepals surrounding inconspicuous, light yellow to pinkish petals. They grow in groups of three on stems that reach up to 1.5 ft. tall. After pollination the flowers turn upward and develop into showy, fluffy seed heads, which are one of the plant's most attractive qualities and source of its common name. *Geum triflorum* makes a lovely addition to sunny gardens where it slowly spreads, creating an attractive groundcover.

Prairie smoke is an interesting groundcover.

Gilia capitata • Polemoniaceae

Bluehead gilia

HABITAT/RANGE Open places, meadows, roadsides, and woodland edges. Southern British Columbia to California. Grows at lower elevations mainly west of the Cascade Mountains extending through the Columbia River Gorge and sporadically into western Idaho.

SEASONAL INTEREST Blue flowers spring to midsummer.

WILDLIFE VALUE Excellent nectar source for butterflies, hummingbirds, bees, and other pollinators. Attracts beneficial insects. Specialist bee host. Moth host plant. Seeds eaten by birds.

CULTIVATION Sun to light shade. Tolerates a wide range of soil types but grows more robustly in well-drained, humus-rich soil. Drought-tolerant annual. Some supplemental water and deadheading can prolong bloom time. Let plants reseed. Sow seed in fall or early spring. Use in sunny gardens, pollinator plantings, and wildflower meadows. Can be crowded out by more vigorous plants.

Bluehead gilia is a slender annual with colorful, globelike clusters of blue flowers. Height depends on site conditions, but the plant can grow to several feet high. This is an attractive and choice plant for pollinator and insectary plantings due to its exceptional ability to attract beneficial insects and provide quality nectar to pollinators. It is easy to grow from seed and a good candidate for seed mixes. Always use ethically sourced seed as wild collection of annual plants can have a big impact on parent populations.

Bluehead gilia has lovely blue flowers that attract butterflies and beneficial insects.

Helenium autumnale • Asteraceae

Common sneezeweed

HABITAT/RANGE Riverbanks, riparian areas, moist meadows, and ditches. Widespread throughout North America. Grows mainly at low elevations on both sides of the Cascade Mountains.

SEASONAL INTEREST Yellow flowers summer into fall.

WILDLIFE VALUE Provides abundant late-season nectar to bees, butterflies, and other pollinators. Butterfly host plant. Many solitary bees specialize on plants in the family Asteraceae. Poisonous. Deer resistant.

CULTIVATION Sun to light shade and moist to wet soil. Does not tolerate dry soil. Will grow in clay soils. Deadhead to encourage more blooms. Grows well in containers. Plant in riparian areas, rain gardens, bioswales, moist gardens, and pollinator plantings. Mulch.

Helenium autumnale is a showy, clump-forming perennial that grows 2 to 4 ft. tall with erect stems and numerous large yellow flowers. The composite flowers are interestingly shaped with a prominent domelike cluster of disc flowers surrounded by a skirt of petal-like ray flowers. This plant likes wet soils and can bloom until frost. It is a great plant for pollinators as it produces nectar late in the season when most other plants have finished blooming. There are many popular cultivars and hybrids that have been derived from this plant; choose plants wisely. Do not be fooled by the name, sneezeweed does not readily agitate allergies.

Helenium autumnale thrives in riparian areas and provides abundant nectar to pollinators in late summer and fall.

Helianthella uniflora • Asteraceae

Rocky Mountain helianthella

HABITAT/RANGE Meadows, rocky slopes, and open woodlands. British Columbia to Nevada and east to Montana at low to high elevations. Grows east of the crest of the Cascade Mountains.

SEASONAL INTEREST Yellow flowers late spring into summer.

WILDLIFE VALUE Attracts bees, butterflies, and other pollinators. Many solitary bees specialize on plants in the family Asteraceae. Seeds eaten by birds and small mammals. Browsed by deer.

CULTIVATION Sun to part shade and well-drained soil. Water to establish. Drought tolerant once established if properly sited. Benefits from some summer water in arid areas. Requires good drainage and is prone to root rot in waterlogged soils. Plant in sunny gardens and pollinator plantings. Mulch lightly.

If you like balsamroot, you will love Rocky Mountain helianthella. This herbaceous perennial has a similar appearance to some of the region's more robust species of balsamroot but grows and reaches flowering age much faster. Stout taproots and a branched woody crown support multiple stems with roughly textured leaves growing 2 to 3 ft. tall or more, each with a single, large, sunflower-like bloom. Give this plant a sunny, prominent position in gardens east of the Cascades where it can attract and feed swallowtail butterflies. A good choice for pollinator plantings on organic farms, orchards, and vineyards.

Helianthella uniflora is a robust perennial with sunflower-like blooms.

Helianthus nuttallii • Asteraceae

Nuttall's sunflower

HABITAT/RANGE Moist meadows and riparian areas at low to mid elevations. Grows mostly in the Rocky Mountains, Great Plains, and across Idaho to eastern Oregon.

SEASONAL INTEREST Yellow flowers midsummer into fall.

WILDLIFE VALUE Late-season food source for bees, butterflies, and other pollinators. Butterfly and moth host plant. Seeds eaten by birds and small mammals. Provides cover.

CULTIVATION Sun to light shade and moist, humus-rich soil. Can grow in clay soils; does not do well in sandy soils. Water to establish and provide supplemental water as needed after establishment. Spreads vigorously by rhizomes in ideal conditions; give it room to grow. Drier soils limit growth. Deadhead. Can be grown in planters but will dominate the container. Mulch.

Nuttall's sunflower is a rhizomatous herbaceous perennial that spreads easily in moist, rich soils that are not compacted. The plant produces multiple stems that grow 3 to 4 ft. tall, sometimes taller, with narrow, roughly textured leaves and showy yellow flowers, which make a nice cut flower. This species of *Helianthus* is not common in Oregon and Washington but is easily grown if given adequate moisture. Being late and long blooming, it is of great value to butterflies and other pollinators and a good plant for insectary and pollinator plantings. The similar Cusick's sunflower (*H. cusickii*) is more common in Oregon and Washington and drought tolerant but not rhizomatous. Common sunflower (*H. annuus*) is a native annual that has been bred to develop the common garden sunflower.

Helianthus nuttallii is a tall, fast-growing perennial that blooms continuously late in the season.

Cow parsnip

Heracleum maximum • Apiaceae

HABITAT/RANGE Moist forests and meadows and riparian areas. Alaska to California and east across North America. Widely distributed throughout the region at low to high elevations.

SEASONAL INTEREST White flowers in summer. Showy seed heads.

WILDLIFE VALUE Important plant for pollinators and beneficial insects. Larval host for multiple species of moths and butterflies, especially important for swallowtail butterflies. Benefits birds. Provides cover. Browsed by deer and elk.

CULTIVATION Sun to shade and moist, humus-rich soil. Grows best in moist shade. Prefers loamy soils but can grow in clay. Water to establish and continue to provide supplemental water as needed. Tolerates soils that dry out by late summer but generally has a low drought tolerance. A large plant; give it ample space. Wear gloves when planting as it can cause skin irritation. Sometimes short-lived or biennial. Mulch.

Grow big with cow parsnip! Huge palmate leaves that can grow up to 1 ft. wide, and generously sized, fragrant, flat-topped umbels of white flowers make this herbaceous perennial a big attraction for both wildlife and gardeners. Useful as an accent or background plant, it can grow 9 to 10 ft. tall and 5 ft. wide in ideal conditions. A good plant for moist woodland gardens, riparian areas, butterfly gardens, and insectary plantings. Be aware that giant hogweed (*Heracleum mantegazzianum*) is an invasive species that looks similar but is significantly larger, reaching 15 ft. tall, with leaves 5 ft. wide.

Cow parsnip is a large, structurally beautiful plant and host for swallowtail butterflies.

Hairy goldaster

Heterotheca villosa • Asteraceae

HABITAT/RANGE Dry, open places, rocky slopes, gravelly streambanks, and shrub-steppe. Widespread throughout the western United States into southern Canada at low to high elevations. Grows primarily east of the Cascade Mountains, with a few scattered populations west of the mountains.

SEASONAL INTEREST Yellow flowers summer into fall.

WILDLIFE VALUE Attracts butterflies, bees, and other pollinators. Many solitary bees specialize on plants in the family Asteraceae. Moth host plant. Birds eat seeds. Deer resistant.

CULTIVATION Sun to light shade and well-drained soil. Will grow in sandy, gravelly, and loamy soils. Water to establish. Drought tolerant once established. Deadhead to encourage more flowers. Plant in dry, sunny pollinator gardens, rock gardens, xeric gardens, and parking strips.

Hairy goldaster is a very drought-tolerant perennial that blooms profusely and continuously through the summer in places that are hot and dry. With few plants able to bloom with such vigor in those conditions, hairy goldaster becomes especially important for native pollinators and butterflies seeking forage. The many stems create rounded clumps up to 20 in. tall from branching, somewhat woody bases with foliage that looks silvery gray due to its hairiness. This plant is very showy when in bloom, producing multitudes of yellow, daisylike flowers. It is great for gardens east of the Cascades.

Hairy goldaster is a prolific bloomer and an excellent plant for xeric pollinator gardens.

Heuchera cylindrica • Saxifragaceae

Roundleaf alumroot

HABITAT/RANGE Rocky slopes and cliffs. British Columbia to Northern California and east to Montana, from mountain foothills to subalpine areas. Grows mainly east of the crest of the Cascade Mountains.

SEASONAL INTEREST Cream to greenish yellow flowers midspring to late summer relative to elevation. Foliage is evergreen in milder climates.

WILDLIFE VALUE Attracts bumble bees, butterflies, hummingbirds, and other pollinators.

CULTIVATION Sun to part shade and rocky, humus-rich, well-drained soil. Grows in full sun in cool, moist sites but needs some shade in hot, dry places. Water to establish and give supplemental water as needed after establishment. Drought tolerant if properly sited. Great for rock gardens and garden borders. Can be grown in planters. Mulch lightly.

Roundleaf alumroot is a lovely perennial with a dense, compact habit that forms clumps of attractive, mostly evergreen, basal leaves. Multiple flowering stems typically grow around 1 ft. tall, sometimes taller, with spikes of cream to greenish yellow flowers that contrast the dark foliage. Gooseberry-leaved alumroot (*Heuchera grossulariifolia*) is similar and grows in the Columbia River Gorge and east of the Cascades but mainly in Idaho. *Heuchera* is a popular genus with gardeners and there are many cultivars and hybrids commonly used in landscaping. Choose plants wisely and favor planting true natives.

Roundleaf alumroot makes a very attractive rock garden plant.

Heuchera micrantha • Saxifragaceae

Smallflower alumroot

HABITAT/RANGE Streambanks, rocky sites, rock crevices, and talus slopes. British Columbia to California and east to Idaho at low to high elevations. Grows primarily west of the Cascade Mountains.

SEASONAL INTEREST White flowers in late spring to late summer relative to elevation. Foliage is evergreen in milder climates.

WILDLIFE VALUE Attracts hummingbirds, bees, butterflies, and other pollinators. Host plant for the moth *Greya politella*.

CULTIVATION Part shade and moist, rocky, humus-rich, well-drained soil. Water to establish and provide supplemental water as needed after establishment. Tolerates seasonally dry conditions if properly sited. Use in shaded rock gardens and woodland gardens. Grows well in planters. Mulch.

Smallflower alumroot is a shade-loving perennial with airy sprays of graceful, delicate white flowers that are showiest when blooming en masse, so plant a few. Red to green, leafless flowering stalks grow 1 to 2 ft. tall from clumps of green to reddish green basal leaves that remain evergreen in milder climates. Smooth alumroot (*Heuchera glabra*) is similar and common from Alaska to northern Oregon, particularly in mountainous regions such as the Cascades and Olympics. *H. micrantha* is the parent plant for many popular garden cultivars and hybrids; choose plants wisely and favor planting true natives.

Heuchera micrantha is a good plant for shady rock gardens west of the Cascades.

Hydrophyllum capitatum • Hydrophyllaceae

Ballhead waterleaf

HABITAT/RANGE Woodlands and seasonally moist, open slopes from southern British Columbia to eastern Oregon and east to the Rocky Mountains at low to high elevations. Grows in and east of the Cascade Mountains.

SEASONAL INTEREST Bluish purple to lavender flowers in early spring, later at higher elevations. Seasonal groundcover.

WILDLIFE VALUE Early-season nectar and pollen source for bees, butterflies, and other pollinators. Especially loved by bumble bees.

CULTIVATION Sun to part shade and seasonally moist, well-drained soil. Water to establish. Drought tolerant once established. Goes dormant after setting seed. Allow soils to dry out when dormant. Plant this ephemeral next to shrubs or summer-blooming perennials that will maintain garden interest throughout the season. Spreads slowly by short rhizomes. Good for dry, partly shaded rock gardens. Mulch.

Ballhead waterleaf is a sweet ephemeral for pollinator gardens east of the Cascades. This herbaceous perennial has attractive, deeply lobed leaves and grows 4 to 15 in. tall. Globelike clusters of bluish purple to lavender flowers grow on stems that, depending on the variation, either extend above or remain below the foliage. There are two variations of *Hydrophyllum capitatum* in the region. *Hydrophyllum capitatum* var. *thompsonii* occurs primarily in the Columbia River Gorge and has flowers that extend above the foliage, while var. *capitatum* has flowers that remain below the leaves. Alpine waterleaf (*H. alpestre*), previously considered a variation of *H. capitatum*, has flowers that bloom at ground level.

Hydrophyllum capitatum var. *thompsonii* has flowers that rise above the foliage.

Hydrophyllum tenuipes • Hydrophyllaceae

Pacific waterleaf

HABITAT/RANGE Moist forests and forest edges at low elevations west of the Cascade Mountains from southwest British Columbia to northwest California.

SEASONAL INTEREST Greenish white flowers spring to early summer. Seasonal groundcover.

WILDLIFE VALUE Attracts bees, butterflies, and other pollinators. Provides cover for small animals.

CULTIVATION Full to part shade and moist, humus-rich soil. Water to establish and continue to provide supplemental water as needed. Spreads by seed and rhizomes. Diminishes in the heat of summer after setting seed; intersperse with evergreen ferns and plants with late-season interest. Perfect for shaded woodland gardens. Mulch.

Pacific waterleaf is a pollinator-friendly groundcover perfect for moist woodlands west of the Cascades. With large, divided and sharply toothed leaves this shade-loving perennial fills the understory of lowland forests in spring. The flowers are held in loose clusters above the leaves and are usually greenish white although some populations may have purple or blue coloration. The plant is hairy and grows 8 to 31 in. tall. Fendler's waterleaf (*Hydrophyllum fendleri*) looks similar but generally grows at higher elevations.

Pacific waterleaf creates a lovely carpet of emerald spring foliage in woodland gardens.

Iliamna rivularis • Malvaceae

Streambank globemallow

HABITAT/RANGE Streambanks, eastside forests, canyons, meadows, and disturbed areas. British Columbia south through Oregon to Nevada and east to Montana at low to high elevations. Grows mainly east of the Cascade Mountains.

SEASONAL INTEREST Pink flowers in summer.

WILDLIFE VALUE Attracts bees, butterflies, hummingbirds, and other pollinators. Butterfly host plant. Provides cover. Browsed by deer and elk. Important food source following a forest fire for many animals.

CULTIVATION Sun to light shade and moist, humus-rich, well-drained soil. Water to establish and continue to provide supplemental water as needed. A large species; give it plenty of space. Great for sunny cottage gardens and pollinator gardens. Mulch.

Another common name for *Iliamna rivularis* is streambank wild hollyhock, and indeed it looks like a hollyhock. The tall, often branched stems boast large palmate leaves and showy racemes of pink, hibiscus-like flowers, making it an attractive and striking garden plant. This herbaceous perennial is long blooming, continuing to flower throughout the summer in moist soils, and usually grows between 2 to 4 ft. tall but can reach 6 ft. tall in ideal conditions. It is adapted to fire, which triggers its seeds to germinate. Seeds can remain viable in the soil for hundreds of years waiting for fire or disturbance. It grows quickly and robustly after a fire but diminishes as forests regrow and shade it out.

Streambank globemallow brings a cottage garden feeling to the landscape.

Ipomopsis aggregata • Polemoniaceae

Scarlet gilia

HABITAT/RANGE Dry meadows, rocky slopes, shrub-steppe, forest openings, and roadsides. Southern British Columbia to California and east to the Rocky Mountains at low to high elevations. Grows primarily east of the crest of the Cascade Mountains.

SEASONAL INTEREST Red flowers late spring to late summer depending on location.

WILDLIFE VALUE Magnet for hummingbirds. Attracts bees, butterflies, and other pollinators. Flowering stalks browsed by elk and deer.

CULTIVATION Sun to part shade and moist to seasonally dry, well-drained soil. Water to establish. Drought tolerant once established. Does well by seed; make sure seed is ethically sourced. Dies after flowering; allow plants to reseed. Flowering stalks are palatable to wildlife, though browsed plants often sprout new stems. Heavy mulch will choke out this plant, which remains low growing for the first year or so.

A truly striking and colorful plant for hummingbird gardens in and east of the Cascades. Scarlet gilia is a biennial or short-lived perennial that for the first year or so remains an attractive, compact basal rosette of leaves until flowering. Herbage is covered in white hairs that give the plant a silvery sheen. A single flowering stalk grows 1 to 3 ft. tall with showy, trumpet-shaped flowers that are scarlet to orange-red with yellow mottling. Flower color varies across its range and may be determined by local pollinators (this plant is used to study pollinator-mediation selection).

Scarlet gilia is brilliantly colored and loved by hummingbirds.

Iris missouriensis • Iridaceae

Rocky Mountain iris

HABITAT/RANGE Moist meadows, streambanks, and vernally wet areas in shrub-steppe and pine forests. Southern British Columbia to California and east beyond the Rocky Mountains. Low to high elevations. Grows east of the Cascade Mountains with populations also in the Puget Sound.

SEASONAL INTEREST Deep to light blue-purple or whitish flowers in late spring to summer.

WILDLIFE VALUE Attracts bees, butterflies, hummingbirds, and other pollinators. Attracts beneficial insects. Deer resistant.

CULTIVATION Sun to part shade and moist to vernally wet soil. Water to establish. Prefers soils that remain consistently moist until flowering but dry out afterwards. Plant in full sun in moist sites; provide afternoon shade in hot, dry sites. Spreads slowly by rhizomes to create clumps. Garden plants can be divided and replanted in fall. Grows well in containers. Mulch lightly.

Irises are favored garden plants and *Iris missouriensis* is no exception. This rhizomatous perennial's flowers are variable but generally have deep to light blue-purple petals (called standards); white to purple, darkly veined sepals (called falls); and yellow throats, which signal pollinators. The large, beardless flowers grow on stems that reach 1 to 2 ft. tall. from clumps of grasslike leaves. This is the only species of native iris found abundantly east of the Cascades in Oregon and Washington and one of the more common and drought tolerant in the region.

Iris missouriensis is a great choice for gardens east of the Cascades.

Iris tenax • Iridaceae

Toughleaf iris

HABITAT/RANGE Prairies, meadows, open forests, and coastal areas. West central Washington to southwestern Oregon with a rare disjunct population in California. Low to mid elevations. Grows west of the Cascade Mountains.

SEASONAL INTEREST Color can vary but typically lavender to purple flowers in midspring to early summer.

WILDLIFE VALUE Attracts bees, butterflies, hummingbirds, and other pollinators. Attracts beneficial insects. Deer resistant.

CULTIVATION Part shade to sun and moist to seasonally dry soil. Can grow in clay. Water to establish and continue to provide supplemental water as needed. Drought tolerant and grows in soils that are dry in summer. Spreads slowly by rhizomes to form clumps; does not respond well to being divided. Can be grown in containers. Plant in rock gardens, garden edges, and partly shaded spots in woodland gardens west of the Cascades. Mulch lightly.

The Pacific Coast is home to multiple species of beardless irises that have been used in plant breeding to create a true rainbow of flower colors and forms. These are known as Pacific Coast irises and *Iris tenax* is one of the prettiest and most common in western Oregon. It is a small but showy perennial growing 8 to 18 in. tall with grasslike leaves and typically purple to lavender, but occasionally yellow, pink, or white flowers. Populations of *I. tenax* are under pressure and becoming increasingly uncommon. Many of the Pacific Coast iris species are rare or endemic; be sure plants are ethically sourced and never dug from the wild.

Iris tenax is a short, beautiful iris found in forests and meadows west of the Cascades.

Lewisia rediviva • Montiaceae

Bitterroot

HABITAT/RANGE Rocky, dry, open sites and shrub-steppe. Southern British Columbia through eastern Washington and Oregon to California and east to the Rocky Mountains at low to fairly high elevations. Grows east of the Cascade Mountains.

SEASONAL INTEREST Pale to deep pink or white flowers spring to early summer relative to elevation.

WILDLIFE VALUE Attracts bees and other pollinators. Foliage eaten by small mammals. Seeds may be eaten by birds.

CULTIVATION Full sun and sharply draining, dry, rocky soil. Very drought tolerant, requires dry summer soils. Goes dormant after setting seed. Water only to establish and do not water when dormant. Can be grown in clay pots or troughs. Plant in rock gardens and dry, sunny areas east of the Cascades. Difficult to grow west of the Cascades.

Bitterroot is a unique perennial with large flowers ranging from pale to deep pink or white with stems so short the blooms appear to erupt from the ground. The basal rosette of succulent, linear leaves withers when plants flower but reemerges in fall and persists through winter. Bitterroot grows only a few inches tall from a branching taproot up to a foot long. The name *rediviva*, meaning "revived," is a nod to its ability to regrow from roots dried for months. The genus *Lewisia* contains showy species that have been used to develop beautiful cultivars and hybrids. Siskiyou lewisia (*L. cotyledon*), from southwestern Oregon and northwestern California, is popular and more tolerant of conditions west of the Cascades. Some species are endemic or uncommon; make sure plants are ethically sourced.

Bitterroot is an unusual low-growing plant with showy flowers.

Lilium columbianum • Liliaceae

Columbia lily

HABITAT/RANGE Meadows, thickets, forest edges and openings, coastal areas, and roadsides. Central British Columbia to northwest California and east through northern Washington to Montana. Low to mid elevations. Grows from the eastern foothills of the Cascade Mountains to the coast, farther east in Washington and British Columbia.

SEASONAL INTEREST Orange flowers in late spring to summer.

WILDLIFE VALUE Attracts bees, butterflies, and hummingbirds. Loved by swallowtail butterflies. Browsed by deer.

CULTIVATION Sun to part shade and moist, well-drained, humus-rich soil. Water to establish. Drought tolerant once established if properly sited but keep soils moist until it finishes flowering. Takes several years from seed to flower. Protect from slugs and deer. Mulch.

Beauty and drama ooze from this incredible wildflower. The nodding flowers are bright orange with maroon spots and have extended anthers so laden with pollen they leave swallowtail butterflies looking like they just came out of a Cheetos factory. Columbia lily is a perennial bulb with erect stems typically growing 2 to 5 ft. tall with whorled leaves and one to twenty flowers. Often called tiger lily, be careful not to confuse *Lilium columbianum* with *L. lancifolium*, the true tiger lily and a native of Asia, which has escaped cultivation and naturalized in New England. There are rare and endemic lilies in the region that look like Columbia lily; make sure plants are ethically sourced, properly identified, and never poached from the wild.

The Pacific Northwest's beauty is accentuated by dramatic wildflowers like Columbia lily.

Lilium washingtonianum • Liliaceae

Washington lily

HABITAT/RANGE Dry slopes, open conifer forests, and chaparral. From Mount Hood in Oregon south through the Cascade Mountains into California. Grows in the Cascade, Klamath, and Sierra Nevada Mountains at low to moderately high elevations.

SEASONAL INTEREST White to pink flowers in summer.

WILDLIFE VALUE Pollinated primarily by moths. Nectar and scent production peak at dusk to correlate with moth activity. Also attracts bees, butterflies, hummingbirds, and other pollinators. Browsed by deer.

CULTIVATION Part shade and well-drained, seasonally dry soil. Water to establish. Drought tolerant once established if properly sited. Takes several years from seed to flower. Likes to grow up through shrubs that help support it and shade its roots. Deer are very attracted to this plant. Mulch.

There are few wildflowers that compare with the showy spectacle of Washington lily. Huge, fragrant, trumpet-shaped flowers are white or pinkish with magenta spots, fading to a darker pink as seed pods begin to develop. These flowers have evolved to attract moths for pollination. A perennial bulb, Washington lily grows as much as 6 ft. tall with whorls of leaves along its sturdy stems. It has a fairly limited range and, despite its name, does not occur naturally in the state of Washington. No matter how prizeworthy this incredibly attractive native plant is, it is essential plants are never poached from the wild and seed collection is done ethically and sparingly.

Washington lily is an incredible wildflower that is pollinated by moths.

Linnaea borealis • Linnaeaceae

Twinflower

HABITAT/RANGE Forest understory, openings, and edges. Widespread from Alaska to California and east across North America at low to high elevations. Circumboreal. Grows on both sides of the Cascade Mountains.

SEASONAL INTEREST Pink flowers early summer to early fall. Evergreen.

WILDLIFE VALUE Attracts bees and other pollinators. Winter food for elk.

CULTIVATION Full to part shade and moist, humus-rich soil. Water to establish. Tolerates fairly dry soils under conifers but supplemental water may be needed in drier sites after establishment. A creeping groundcover that does not spread aggressively, it may take many years for this plant to start spreading and blooming from seed. Mulch lightly. Heavy applications of mulch may choke out this low-growing plant, yet it likes to grow in soils amply amended with broken down leaf litter.

Twinflower is a lovely evergreen groundcover for moist forest gardens. It has thin woody runners that vine their way under trees and shrubs forming a carpet of glossy green leaves. Fragrant, light pink, bell-shaped flowers bloom in pairs atop flowering stems that grow 4 in. tall, lighting up the forest floor when blooming en masse. A widely distributed plant appropriate for gardens on either side of the Cascades that can give it a cool, moist, shady spot.

Twinflower creeps through forest understories, creating a shiny evergreen groundcover.

Linum lewisii • Linaceae

Wild blue flax

HABITAT/RANGE Open areas, meadows, grasslands, forest openings, and roadsides. Widespread throughout much of western North America at low to high elevations. Grows mainly east of the crest of the Cascade Mountains.

SEASONAL INTEREST Blue flowers late spring to summer.

WILDLIFE VALUE Attracts bees, butterflies, and other pollinators. Seeds eaten by birds and other animals. Tender spring foliage and seedlings browsed by deer and other wildlife, but flowering plants have fibrous stems that tend to be avoided.

CULTIVATION Sun to light shade and well-drained soil. Water to establish. Drought tolerant once established. Can tolerate poor soils with good drainage. Reseeds readily and looks beautiful in large swaths. Allow plants to reseed or save garden seed to resow as it is a short-lived perennial. Cut back before seed ripens to curtail spreading if needed. Use in wildflower meadows, pollinator plantings, and parking strips. Does well in planters.

Wild blue flax can quickly fill a sunny garden with waves of blue blooms. Its many branching stems grow to around 2 ft. tall from woody crowns, bending gracefully outward from the center with short, narrow leaves and multiple blue flowers at the tips. It grows readily from seed and is popular in seed mixes. Be aware there are a few similar nonnative species of *Linum* that have escaped birdseed mixes and cultivation and are now naturalized in the region. A European species, common flax (*L. usitatissimum*), is commonly cultivated for food and fiber.

Wild blue flax produces copious beautiful blue flowers.

Lithophragma parviflorum • Saxifragaceae

Smallflower prairiestar

HABITAT/RANGE Meadows, coastal bluffs, rocky balds, shrub-steppe, grasslands, oak woodlands, and forest openings. British Columbia to California and east to the Great Plains at low to fairly high elevations. Grows on both sides of the Cascade Mountains.

SEASONAL INTEREST Pink to white flowers in spring to early summer relative to elevation.

WILDLIFE VALUE Attracts native pollinators and the pollinating floral parasite *Greya politella*, a moth that lays eggs in the flowers they pollinate.

CULTIVATION Sun to part shade and well-drained, vernally moist soil. Water to establish. Drought tolerant once established and prefers dry summers. Dies back after setting seed. Place near plants that will keep interest in the garden throughout the season but not overwhelm this low-growing perennial. Great for rock gardens and xeric gardens.

Smallflower prairiestar is a sweet ephemeral perennial that forms small bulblets along slender rhizomes that grow shallowly in the soil. The small but showy flowers are pink to white and featherlike, blooming in clusters on stems that grow from 4 in. to just over 1 ft. tall depending on site conditions. The small, hairy, deeply lobed basal leaves can be described as pretty darn cute. Darling as they are, these plants grow in some pretty tough places where they avoid heat and drought by growing robustly during the wet spring and going dormant by summer. There are other species of *Lithophragma* in the region that also make lovely garden plants.

Smallfower prairiestar is a rugged and resilient plant that looks delicate and dainty.

Lomatium dissectum • Apiaceae

Fernleaf desert parsley

HABITAT/RANGE Open areas, rocky slopes, and open woodlands. Southern British Columbia to California at low to high elevations. Disjunct in Idaho. Red listed in British Columbia. Grows on both sides of the Cascade Mountains.

SEASONAL INTEREST Brownish purple or yellow flowers midspring to early summer relative to elevation.

WILDLIFE VALUE A great plant for pollinators and beneficial insects. Specialist bee host. Swallowtail butterfly host plant. Browsed by deer and elk.

CULTIVATION Sun to light shade and well-drained, seasonally dry soil. Water to establish. Drought tolerant once established. Goes dormant after setting seed. Do not water when dormant and do not plant in gardens with continual irrigation. Give this large species plenty of room but place near plants that will maintain garden interest after it goes dormant. Spreads by seed; cut back seed heads before ripe to curtail spreading if needed. Plant in xeric butterfly gardens.

Fernleaf desert parsley is a large taprooted perennial with attractive fernlike leaves and umbels of tiny brownish purple, sometimes yellow, flowers in globelike clusters on stout stems that can grow 5 ft. tall. Height and form make this a striking plant. *Lomatium multifidum* is nearly identical and until recently was classified as a variation of *L. dissectum*. It grows in a similar range but chiefly east of the Cascades extending to the Rocky Mountains. Both species make bold additions to butterfly gardens.

Lomatium dissectum is a large perennial that makes a bold statement in sunny, xeric butterfly gardens.

Lomatium nudicaule • Apiaceae

Barestem desert parsley

HABITAT/RANGE Open areas, meadows, rocky slopes, shrub-steppe, oak woodlands, and open pine forests. Widespread from southern British Columbia to California and east to Idaho and Utah. Low to mid elevations. Grows on both sides of the Cascade Mountains.

SEASONAL INTEREST Yellow flowers spring to midsummer relative to elevation. Structural interest lasts through winter.

WILDLIFE VALUE Swallowtail butterfly host plant. Attracts bees, butterflies, and other pollinators. Specialist bee host. Plants eaten by small mammals.

CULTIVATION Sun to light shade and well-drained, seasonally dry soil. Can grow in sandy soils. Water only to establish. Drought tolerant once established. Plant in sunny, dry meadows, rock gardens, xeric gardens, and pollinator plantings. Mulch lightly.

Barestem desert parsley is an aromatic herbaceous perennial with compound umbels of yellow flowers. The globelike flower clusters are colorfully compact as they begin blooming close to the ground but continue to expand and heighten on sturdy stems that can grow up to 30 in. tall. When fully developed the large umbels provide a structurally striking garden accent that often persists through winter. These plants have strong taproots and basal leaves with ovate, bluish green, somewhat leathery leaflets. The leaves are edible and spicy with a distinct celery-like flavor. A great plant for sunny xeric gardens that is beneficial to pollinators.

Lomatium nudicaule looks beautiful blooming en masse in dry meadows and its striking structure leaves a lasting impression in sunny xeric gardens.

Lomatium papilioniferum • Apiaceae

Butterfly-bearing desert parsley

HABITAT/RANGE Sunny, dry, open areas, rocky slopes, and outcroppings. Grows at low to mid elevations east of the Cascade Mountains in Oregon, Washington, and Idaho.

SEASONAL INTEREST Yellow flowers in early spring to early summer relative to elevation.

WILDLIFE VALUE Larval host for swallowtail butterflies, particularly important for the Indra swallowtail. Early-season nectar source for bees and other pollinators. Specialist bee host. Birds glean insects from plants. Eaten by elk, deer, and other mammals.

CULTIVATION Full sun and rocky, well-drained, seasonally dry soil. Goes dormant after setting seed. Water to establish but let soils dry out after it goes dormant. Drought tolerant once established. Do not plant in irrigated gardens. Best grown east of the Cascades. Great for sunny xeric, butterfly, and rock gardens. Mulch lightly.

Papilio is the genus name for swallowtail butterflies, which utilize species of *Lomatium* as larval hosts and adult nectar sources, *Lomatium papilioniferum* being one of them. This is a strongly aromatic herbaceous perennial whose scent signals spring to those who know it. From stout taproots grow mounds of attractive dill-like foliage and compound umbels of yellow flowers on stems that are 6 to 20 in. tall, sometimes taller. Spring gold (*L. urticulatum*) is similar and more appropriate for gardeners west of the Cascades. There are many species of *Lomatium* found in the Pacific Northwest that make lovely additions to butterfly and pollinator gardens. Until recently, *L. papilioniferum* was classified as *L. grayi*.

Lomatium papilioniferum is pungent and early blooming.

Lupinus latifolius • Fabaceae

Broadleaf lupine

HABITAT/RANGE A variable plant found in many different habitats including meadows, rocky areas, riparian areas, open forests, and roadsides in western North America. Low to high elevations. Grows on both sides of the Cascade Mountains.

SEASONAL INTEREST Bluish purple to pink flowers in spring to summer relative to elevation.

WILDLIFE VALUE Attracts bees, bumble bees, butterflies, humming-birds, and other pollinators. Specialist bee host. Host plant for multiple species of butterflies. Attracts beneficial insects. Seeds eaten by birds and small mammals. Deer resistant.

CULTIVATION Sun to part shade and moist to seasonally dry, well-drained soil. A variable species adapted to a wide range of conditions; use plants propagated from locally sourced seed to ensure they are adapted to your climate. Water to establish and provide supplemental water as needed after establishment. Drought tolerant if properly sited. May be easier to establish by seed. Mulch lightly.

Broadleaf lupine creates vivid swaths of color in meadows and open forests and can bring a riot of color and pollinators to the garden as well. Bluish purple to pink spikes of pealike flowers grow 1 to 4 ft. tall over mounds of attractive green to silvery green foliage. Below the soil the roots of this leguminous perennial work to fix nitrogen, making it an important plant for rehabilitating disturbed and depleted soils. An attractive choice for perennial beds, butterfly gardens, and wildflower meadows. Hybridizes with other species of lupine.

A beautiful patch of broadleaf lupine.

Lupinus leucophyllus • Fabaceae

Velvet lupine

HABITAT/RANGE Open areas, grasslands, shrub-steppe, and dry woodlands. Southern British Columbia to Northern California and east to Montana at low to mid elevations. Grows mainly east of the Cascade Mountains.

SEASONAL INTEREST Purple to whitish flowers in late spring to late summer. Silvery foliage.

WILDLIFE VALUE Attracts bees, bumble bees, hummingbirds, butter-flies, and other pollinators. Attracts beneficial insects. Specialist bee host. Seeds eaten by birds and small mammals. Deer resistant.

CULTIVATION Sun to light shade and well-drained soils. Prefers rocky soils that are moist in spring and dry in summer. Water to establish. Drought tolerant once established. Mulch lightly.

Velvet lupine is a unique and robust species with dense, showy spikes of pealike flowers that range in color from bluish purple to pink or whitish. One of the most stunning qualities of this plant is its silvery haired foliage, which is velvety to the touch. This is a very drought-tolerant herbaceous perennial that grows up to 3 ft. tall in sunny sites east of the Cascades where it makes an attractive addition to pollinator gardens. Silky lupine (*Lupinus sericeus*) and sulfur lupine (*L. sulphureus*), which has yellow flowers, are also drought tolerant and fairly widespread east of the Cascades. Like other leguminous plants, lupines fix nitrogen and are useful for rehabilitating soils.

Velvet lupine is a drought-tolerant perennial with dramatic beauty.

Lupinus polyphyllus • Fabaceae

Bigleaf lupine

HABITAT/RANGE Moist meadows and riparian areas. Widespread in western North America with many intergrading variations. Low to high elevations. Grows on both sides of the Cascade Mountains.

SEASONAL INTEREST Bluish purple flowers in late spring to summer. Showy, pealike seed pods.

WILDLIFE VALUE Attracts bees, bumble bees, hummingbirds, butterflies, and other pollinators. Butterfly host plant. Attracts beneficial insects. Specialist bee host. Seeds eaten by birds and small mammals. Deer resistant.

CULTIVATION Sun to part shade and moist soil. Water to establish and continue to provide supplemental water as needed. Plant in wet meadows, riparian areas, moist butterfly gardens, rain gardens, and irrigated flower beds. Spreads vigorously by seed in ideal conditions; remove seed pods before ripe to curtail spreading if needed. Mulch.

Gardeners may not realize the predominant parent plant for many of the most popular cultivated varieties of lupines is this robust native species. *Lupinus polyphyllus* is a bold herbaceous perennial that can grow 3 to 5 ft. tall with spikes of bluish purple, pealike flowers and attractive, palmately divided leaves. Nitrogen-fixing roots, an affinity for moist soils, and an attractiveness to butterflies, pollinators, and wildlife make this a useful species for wetland restoration. Streambank lupine (*L. rivularis*) is another good species for wetland plantings. There are many lupines native to the region; plant the species and variations that grow near you.

Bigleaf lupine is a vigorous and showy species that has been used in plant breeding to produce some of the most commonly cultivated varieties of lupines.

Lysichiton americanus • Araceae

Skunk cabbage

HABITAT/RANGE Wetlands, bogs, swamps, wet forests, coastal areas, and riparian areas. Alaska to California and east to Montana at low to mid elevations. Grows on both sides of the Cascade Mountains.

SEASONAL INTEREST Greenish yellow flowers surrounded by a showy yellow spathe bloom in late winter to early summer depending on location. Large leaves.

WILDLIFE VALUE Attracts pollinating and beneficial insects. Mainly pollinated by rove beetles, which forage for pollen and mate in the inflorescences. Plants eaten by a variety of animals including elk, deer, bear, and squirrels.

CULTIVATION Shade to part sun and wet, humus-rich soil. Tolerates a variety of soil types but needs consistent moisture. Slow to develop but long lived. Plant in wetlands, riparian areas, and soggy spots.

This Pacific Northwest icon is distinctive in size, scent, and structure. Blooming as soon as the snow melts, skunk cabbage makes a showy spring appearance with tight clusters of tiny flowers in a clublike spadix that grows to about 1 ft. tall surrounded by a large, bright yellow, capelike bract called a spathe. The flowers often emerge before the leaves and in time produce densely packed green to reddish fruits. The broad leaves are huge, eventually reaching as much as 5 ft. long. Both the flowers and foliage lend a lush, tropical feeling to shady wetlands. A choice plant for wet sites despite its namesake smell. This plant has escaped cultivation in Europe where it is considered invasive.

Skunk cabbage is a beacon in wet, shady lowlands.

Maianthemum dilatatum • Asparagaceae

False lily-of-the-valley

HABITAT/RANGE Moist forests, shady streambanks, and coastal areas from Alaska to California. Grows at low to mid elevations from the coast to Cascade foothills, occurring occasionally east of the Cascade Mountains and farther inland to northern Idaho.

SEASONAL INTEREST White flowers in late spring to early summer. Speckled red berries. Seasonal groundcover.

WILDLIFE VALUE Attracts pollinating and beneficial insects. Berries eaten by birds and other wildlife.

CULTIVATION Full to part shade and moist, humus-rich soil. Water to establish. Tolerates seasonally dry conditions once established if properly sited but maintaining moderate soil moisture will keep plants looking lush. A vigorous groundcover that spreads by rhizomes and may overwhelm more delicate plants. Divisions of garden plants can be transplanted in spring or fall. Mulch.

False lily-of-the-valley is an attractive shade-loving groundcover that is easily grown in moist, wooded areas. This herbaceous perennial forms dense carpets of heart-shaped leaves and is vigorous enough to keep low-growing invasive weeds at bay. Above the shiny and strongly veined foliage, spikes of white flowers grow 4 to 15 in. tall and develop into greenish, cream-colored berries speckled with red that become thoroughly translucent red when ripe. These berries can provide lasting interest into the fall although the leaves may start to yellow and wither before fruits are ripe. An excellent plant for coastal and moist shady gardens.

False lily-of-the-valley fills the understory of western coastal forests.

Maianthemum racemosum • Asparagaceae

Large false Solomon's seal

HABITAT/RANGE Forests, woodlands, moist meadows, and streambanks at low to mid elevations. Widespread in North America. Grows on both sides of the Cascade Mountains.

SEASONAL INTEREST White flowers in midspring to midsummer depending on location. Red berries.

WILDLIFE VALUE Attracts small bees, beetles, and other pollinators. Attracts beneficial insects. Fruits eaten by birds and small mammals. Occasionally browsed by deer.

CULTIVATION Part shade and moist to seasonally dry, humus-rich soils. Water to establish. Drought tolerant once established if properly sited but maintaining moderate soil moisture will keep plants looking lush. Grows well in containers. Mulch.

Large false Solomon's seal is a handsome herbaceous perennial for partly shaded gardens with soils amended with organic matter. It is rhizomatous but has a clumping habit, growing 2 to 3 ft. tall with pointed, oblong, conspicuously veined leaves alternating along arching stems that terminate in showy plumes of small, fragrant, white flowers. It produces green and red speckled berries that turn translucent red when ripe. The flowers are visited by a wide array of pollinators and the attractive fruits are relished by birds, making this a choice plant for woodland gardens striving for a beautiful blend of form and function.

Large false Solomon's seal is a beautiful and charismatic shade-loving perennial.

Maianthemum stellatum • Asparagaceae

Star-flowered Solomon's seal

HABITAT/RANGE Forests, woodlands, streambanks, and rocky hillsides at low to high elevations. Widespread across most of North America. Grows on both sides of the Cascade Mountains.

SEASONAL INTEREST White flowers in late spring to summer relative to elevation. Green to red berries.

WILDLIFE VALUE Attracts bees and other pollinators. Attracts beneficial insects. Fruits eaten by small mammals and birds. Dense patches provide cover for small wildlife. Browsed by elk and deer.

CULTIVATION Part sun to shade and moist but well-drained, humus-rich soil. Water to establish. Tolerates seasonally dry conditions once established if properly sited but maintaining moderate soil moisture will keep plants looking lush. Occurs in a variety of habitats across its range but grows best in partly shady sites with soils rich in organic matter. Spreads by rhizomes. Divisions from garden plants can be transplanted in spring or fall. Mulch.

Star-flowered Solomon's seal grows up to 2 ft. tall and has an altogether pleasing appearance and structure. Its stems rise from stout rhizomes and zigzag between pointed, parallel-veined leaves, culminating in five to fifteen small, white, star-shaped flowers. After pollination greenish berries with a reddish stripe develop, turning completely dark red when ripe. This herbaceous perennial makes for a lovely groundcover when growing densely but is also attractive popping up here and there among other shade-loving plants. A great plant for woodland gardens.

Star-flowered Solomon's seal is a lovely rhizomatous woodland perennial.

Mertensia paniculata • Boraginaceae

Tall bluebells

HABITAT/RANGE Streambanks, wet meadows, open woods, and talus slopes. Alaska to Oregon and east to Montana, with one variation of the species extending to the Great Lakes. Low to high elevations. Grows on both sides of the Cascade Mountains.

SEASONAL INTEREST Blue flowers in late spring to late summer depending on location.

WILDLIFE VALUE Attracts bees, butterflies, and other pollinators. Specialist bee host. Browsed by elk and deer.

CULTIVATION Sun to shade and moist, humus-rich soil. Does best with some shade at lower elevations. Water to establish and continue to provide supplemental water as needed. Mulch.

A pleasing plant reminiscent of comfrey with erect to slightly arching leafy stems that grow to about 3 ft. tall in ideal conditions, and clusters of pendulous, blue, bell-shaped flowers. The foliage of this herbaceous perennial often has a glaucous hue that complements the flower color nicely. Tall bluebells is prolific in northern latitudes of western North America. In the southern part of its range it tends to take refuge in cool mountain soils. If you can give this plant adequate moisture and a relatively cool site it will grow beautifully. Plant in moist gardens with afternoon shade. There are many species of *Mertensia* native to the Pacific Northwest and the genus is particularly diverse in the Rocky Mountains.

Mertensia paniculata is a tall perennial with a clumping habit and blue, bell-shaped flowers.

Monardella odoratissima • Lamiaceae

Mountain monardella

HABITAT/RANGE Open, rocky areas from Washington to California and east to Idaho at low to fairly high elevations. Grows mainly east of the Cascade Mountains, farther west in southern Oregon and California.

SEASONAL INTEREST Pinkish purple to white flowers in summer.

WILDLIFE VALUE Great for pollinators, especially loved by bumble bees and butterflies. Monarch butterfly nectar plant. Attracts beneficial insects. Specialist bee host. Deer resistant.

CULTIVATION Sun to light shade and rocky, well-drained soil. Water to establish. Do not overwater, let soil dry out between waterings. Drought tolerant once established. Great for rock gardens and sunny, xeric pollinator plantings.

Mountain monardella satisfies the senses of both sight and smell with its attractive, blue-green to green, mounding foliage and profusions of pink-purple flowers that emit a minty fragrance. This herbaceous, somewhat woody, perennial is low growing, reaching as much as 20 in. tall, and thrives in sunny sites with good drainage and low competition from other plants. A great pollinator plant that attracts bumble bees so eager for the nectar they may try to pry flowers open before they are fully unfurled. Leaves can be used to make a minty tea and dried seed heads make an excellent potpourri. This genus is particularly diverse in California with a few other species ranging into southwestern Oregon, such as coyote mint (*Monardella villosa*), which is a bit taller, attracts hummingbirds, and is a host plant for many species of moths.

Mountain monardella has a minty fragrance, mounding habit, and flowers loved by bumble bees and butterflies.

Nothochelone nemorosa • Plantaginaceae

Woodland beardtongue

HABITAT/RANGE Forests and rocky slopes from southwestern British Columbia to northwestern California. Low to high elevations. Grows mainly in the Cascade Mountains and west to the coast, occasionally in the mountains of eastern Washington and Oregon.

SEASONAL INTEREST Rose-purple flowers summer to early fall.

WILDLIFE VALUE Attracts hummingbirds, bees, and other pollinators.

CULTIVATION Part shade to sun and moist to seasonally dry, well-drained soil. Water to establish. Tolerates seasonally dry conditions once established if properly sited but occasional summer water will keep plants looking lush. A woodland plant that prefers part shade and soils amended with organic matter but not overly rich or saturated. Mulch.

Woodland beardtongue is an eye-catching plant with large, tubular, rose-purple flowers that beckon to bees and hummingbirds. Erect or arching stems grow to 2.5 ft. with coarsely toothed leaves and open clusters of flowers at the tips. In cool mountain forests and rocky slopes this robust herbaceous perennial can potentially bloom until early fall providing a late-season nectar source for pollinators. At lower elevations or in warmer environments it will likely finish blooming by late summer. Similar to and once classified as a penstemon, it is the only species of *Nothochelone* in the world.

Woodland beardtongue has large tubular flowers that attract hummingbirds.

Olsynium douglasii • Iridaceae

Grass widow

HABITAT/RANGE Meadows, open areas, rocky bluffs, forest openings, and oak woodlands. Southern British Columbia to Northern California and east to Idaho at low to mid elevations. Grows on both sides of the Cascade Mountains, though mainly to the east.

SEASONAL INTEREST Magenta, sometimes white or striped, flowers in late winter to early summer relative to elevation.

WILDLIFE VALUE Early-season nectar and pollen source for bees, including bumble bees and mason bees, as well as other pollinators.

CULTIVATION Sun to light shade and vernally moist soil. A spring ephemeral that goes dormant after setting seed. Water to establish but let soils dry out in summer when dormant. Drought tolerant once established. Let plants self-sow or collect and sow seed from garden plants to increase population. Plant in rock gardens, wildflower meadows, and sunny, xeric gardens. Mulch lightly.

Grass widows are one of the earliest blooming wildflowers in the region, sometimes blooming in late winter if the weather is mild. They flourish in open areas where soils are moist in spring but dry in summer and grow in clumps about 1 ft. tall with grasslike leaves and large flowers that are typically deep magenta but sometimes white or striped. Use this herbaceous perennial in place of crocus. Previously classified as *Sisyrinchium douglasii*, it is now the only species of *Olsynium* in North America, with two variations found in the region.

Grass widows are lovely early-blooming wildflowers.

Opuntia spp. • Cactaceae

Pricklypear

HABITAT/RANGE Sunny, dry, rocky areas from British Columbia south and east to various parts of North America. Low to mid elevations depending on the species. *Opuntia fragilis* and *O. ×columbiana* occur east of the Cascade Mountains in Washington and Oregon, with *O. fragilis* also growing in the Puget Sound.

SEASONAL INTEREST Yellow to peach or reddish flowers late spring to summer. Evergreen.

WILDLIFE VALUE Attracts a variety of pollinators. Fruits eaten by birds and mammals. Deer resistant.

CULTIVATION Full sun and dry, rocky soil with sharp drainage. Very drought tolerant. Water only to establish. Use chopsticks to avoid spines and move detached pads from garden plants to new areas. Place bottom tip of pad in soil to propagate. Do not take plants from the wild. Plant in rock gardens and sunny, xeric sites. Make sure site is well weeded before planting as weeding around these plants is difficult. Can be grown in containers. Gravel mulch.

People don't think of the Pacific Northwest as cactus country, but a few call the region home and pricklypear is the most common. Pricklypears are low growing, reaching about 1 ft. tall, and form mats of spiny, flattened pads. Their large, showy flowers have many yellow to peach or reddish petals and attract pollinators. Brittle pricklypear (*O. fragilis*) is widespread and grows farther north than any other species of cactus, occurring far into British Columbia. In some areas these plants flower infrequently yet still provide an evergreen groundcover with needle-sharp spines that will effectively discourage any foot traffic.

Pricklypear is a low-growing cactus.

Oxalis oregana • Oxalidaceae

Oregon wood-sorrel

HABITAT/RANGE Moist forests from the west slope of the Cascade Mountains to the coast. British Columbia to California. Mainly at low elevations.

SEASONAL INTEREST White to pink flowers in spring to summer. Seasonal groundcover.

WILDLIFE VALUE Attracts bees, butterflies, and other pollinators. Seeds eaten by birds and small mammals, which also browse foliage.

CULTIVATION Full to part shade and moist, acidic, humus-rich soil. Water to establish and continue to provide supplemental water as needed. Tolerates seasonally dry conditions once established if properly sited. Spreads but shallow rhizomes are easily controlled. Use as a groundcover in shady, moist gardens west of the Cascades. Mulch.

Oregon wood-sorrel's emerald, shamrock-shaped leaves carpeting the floor of westside forests is a familiar sight. Growing from shallow rhizomes, these juicy leaves are edible and have a tart flavor but should only be consumed in small quantities. This delicate herbaceous perennial has white to pink, red-veined flowers that grow to about 6 in. tall. Direct sunlight causes the leaves to collapse and fold downward; they resurrect themselves when shadier conditions return. This also happens at night making it seem as if the plants are sleeping. This sweet understory plant is a vigorous spreader in moist, shady areas with soils rich in organic matter.

Oregon wood-sorrel is a charming groundcover found in moist forests.

Penstemon davidsonii • Plantaginaceae

Davidson's penstemon

HABITAT/RANGE Open, rocky sites in mountainous areas from British Columbia to California and northwest Nevada. Mid to high elevations. Found on both sides of the Cascade Mountains.

SEASONAL INTEREST Blue-lavender to rose-purple flowers in spring to late summer relative to elevation. Evergreen.

WILDLIFE VALUE Attracts bumble bees, bees, hummingbirds, and other pollinators. Specialist bee host. Larval host for checkerspot butterflies. Attracts beneficial insects. Seeds may be eaten by birds. Deer resistant.

CULTIVATION Sun to light shade and rocky, well-drained soil. Do not plant in excessively rich soil. Water to establish. Drought tolerant once established. A choice rock garden plant. Can be grown in containers. Hybridizes with other penstemons in the subgenus *Dasanthera*.

Penstemon is a genus endemic to Central and North America that contains hundreds of species, all of which are important to local pollinators, especially bees. Considered some of the region's most beautiful wildflowers, they vary greatly in flower color and size, as well as leaf type and growth habit. Found chiefly in dry and mountainous areas in and east of the Cascades, some species, like Davidson's penstemon, also grow west of the mountain range. *Penstemon davidsonii* is a woody, low-growing, mat-forming evergreen perennial that carpets itself with large, blue-lavender to rose-purple, tubular flowers. Growing only 4 in. tall, it makes a gorgeous addition to the edge of rock gardens, pollinator gardens, and well-drained sites that mimic the rocky conditions in which it grows naturally.

Penstemon davidsonii is a low-growing evergreen perennial that puts on a colorful show with its tubular flowers.

Penstemon euglaucus • Plantaginaceae

Glaucous penstemon

HABITAT/RANGE Open slopes and conifer forests. Found primarily at mid elevations in the Cascade Mountains from Mount Adams in Washington south through central Oregon.

SEASONAL INTEREST Blue-purple flowers in late spring to summer, occasionally again in fall. Semievergreen.

WILDLIFE VALUE Attracts bees, hummingbirds, butterflies, and other pollinators. Specialist bee host. Larval host for checkerspot butterflies. Attracts beneficial insects. Seeds may be eaten by birds. Deer resistant.

CULTIVATION Sun to part shade and well-drained soil. Water to establish. Drought tolerant once established if properly sited. A relatively shade-tolerant penstemon, it prefers at least partial sun and doesn't like deep, moist shade. Plant in bee gardens, rock gardens, and well-drained perennial beds. Can be grown in containers.

Glaucous penstemon has grayish blue-green foliage that contrasts its blue-purple tubular flowers beautifully. The large, spoon-shaped basal leaves usually persist through winter, often turning purple. Erect flowering stems with narrower leaves grow 1 to 2 ft. tall and bear clusters of flowers that attract pollinators. This perennial forms a gorgeous groundcover in gardens, but it does have a fairly limited natural range. Rydberg's penstemon (*Penstemon rydbergii*) is similar and more widespread, growing in moist to dryish sites mostly east of the Cascades. Fine-toothed penstemon (*P. subserratus*), which grows primarily on the eastern flanks of the Cascades in Washington and northern Oregon, is another great pollinator plant that is lovely in gardens. There are many species of *Penstemon* native to the region; find out which ones grow in your area.

Glaucous penstemon is a beautiful, relatively shade-tolerant penstemon.

Penstemon fruticosus • Plantaginaceae

Shrubby penstemon

HABITAT/RANGE Rocky sites and open woodlands at moderately low to high elevations in the mountains from British Columbia to Oregon and east to Montana. Grows primarily east of the crest of the Cascade Mountains.

SEASONAL INTEREST Blue-lavender to lilac-purple flowers in late spring to late summer relative to elevation. Evergreen.

WILDLIFE VALUE Attracts bumble bees, bees, hummingbirds, butterflies, and other pollinators. Specialist bee host. Checkerspot butterfly host plant. Attracts beneficial insects. Seeds may be eaten by birds. Deer resistant.

CULTIVATION Sun to light shade and well-drained soil. Do not plant in excessively rich soil. Water to establish. Drought tolerant once established if properly sited. Some shade may be needed in hot, lowland environments. Grows well in containers. Hybridizes easily with other penstemons in the subgenus *Dasanthera*.

Penstemon fruticosus is a woody perennial that grows up to 16 in. tall but is usually shorter. It forms attractive mats of shiny evergreen foliage that may turn reddish in winter. Its showy tubular flowers come in various shades of purple and have woolly anthers, a hairy lower lip, and a bearded staminode (sterile stamen that all penstemons have), which helps ensure pollen distribution and dispersal by bees. This is a lovely perennial for the rock garden where it spills nicely over the edges of rock walls. Gardeners west of the Cascades will want to plant Cardwell's penstemon (*P. cardwellii*), which is similar and also heat and drought tolerant but adapted to moister westside environments.

Penstemon fruticosus is an attractive, evergreen, shrubby perennial perfect for rock gardens.

Penstemon richardsonii • Plantaginaceae

Richardson's penstemon

HABITAT/RANGE Rocky sites and cliffs at low to moderately low elevations from southern British Columbia to central Oregon. Grows mainly east of the crest of the Cascade Mountains.

SEASONAL INTEREST Pink flowers early to late summer, possibly into early fall.

WILDLIFE VALUE Late-season nectar source. Attracts bumble bees, bees, hummingbirds, butterflies, and other pollinators. Specialist bee host. Butterfly host plant. Attracts beneficial insects. Seeds may be eaten by birds. Deer resistant.

CULTIVATION Sun to part shade and rocky, well-drained soil. Water to establish. Drought tolerant once established if properly sited. Plant in sunny rock gardens, pollinator plantings, and parking strips. Can be grown in containers. Responds well to a light gravel mulch.

Penstemon richardsonii is a beautiful and fairly adaptable species. Often found growing out of cliff faces and rocks in hot, dry areas, it also thrives on gravelly river islands. While sometimes finnicky, if given good drainage it can thrive in the garden as well. It is a woody perennial with deeply toothed leaves and stems reaching 8 to 31 in. tall. The pink flowers are large and showy, and an important late-season nectar source for pollinators like bumble bees and hummingbirds. Also known as cut-leaf penstemon, it is occasionally short-lived but usually grows vigorously for many years, making a colorful statement in xeric gardens.

Penstemon richardsonii has showy, pink, tubular flowers that are loved by bumble bees and hummingbirds.

Penstemon serrulatus • Plantaginaceae

Cascade penstemon

HABITAT/RANGE Moist areas, meadows, seeps, streambanks, and forest openings at low to mid elevations from British Columbia to central Oregon. Grows mainly in and west of the Cascade Mountains.

SEASONAL INTEREST Purple to blue flowers in early to late summer relative to elevation. Fall color.

WILDLIFE VALUE Attracts bees, butterflies, hummingbirds, and other pollinators. Specialist bee host. Checkerspot butterfly host plant. Attracts beneficial insects. Seeds may be eaten by birds. Deer resistant.

CULTIVATION Sun to part shade and moist to seasonally moist, well-drained, rocky to humus-rich soils. Water to establish and continue to provide supplemental water as needed after establishment. Drought tolerant if properly sited. Plant in pollinator gardens west of the Cascades. Will grow east of the Cascades and can be heat and drought tolerant if given a partly shaded site. Can be grown in containers. Mulch.

Most penstemons thrive in dry, sunny sites like those found east of the Cascades, but *Penstemon serrulatus* breaks this stereotype, preferring moist, and even shaded, soils on the wet side of the mountains. This herbaceous perennial is an attractive summer bloomer that entices pollinators like hummingbirds. Growing to around 2 ft. tall, it has whorl-like clusters of purple to blue tubular flowers on erect stems with serrated leaves that can turn lovely reddish colors in fall. Broad-leaved penstemon (*P. ovatus*) is taller and also grows mainly in and west of the Cascades.

Penstemon serrulatus is unlike most species of penstemon in that it can grow in moist soils.

Penstemon speciosus • Plantaginaceae

Showy penstemon

HABITAT/RANGE Dry, sunny areas, shrub-steppe, and open juniper and pine forests from Washington to California and east to Idaho and Nevada. Low to high elevations. Grows east of the Cascade Mountains.

SEASONAL INTEREST Blue to purple flowers late spring to mid-summer relative to elevation.

WILDLIFE VALUE Attracts bees, butterflies, hummingbirds, and other pollinators. Specialist bee host. Butterfly host plant. Attracts beneficial insects. Seeds may be eaten by birds. Deer resistant.

CULTIVATION Sun to light shade and well-drained soil. Can grow in sandy soil. Do not plant in excessively rich soil. Water to establish. Drought tolerant once established. Plant in rock gardens and sunny, xeric pollinator gardens.

Penstemon speciosus puts on a colorful show in sunny, dry places east of the Cascades with its large, blue to purplish, tubular flowers. Multiple stout stems grow to about 2.5 ft. tall with narrow, linear leaves and more elliptic leaves clustered around the base. Showy penstemon creates an oasis for pollinators and insects, which in turn attract and feed birds. This drought-tolerant herbaceous perennial requires little to no water after establishment and makes a vibrant statement in xeric gardens. There are many species of *Penstemon* found in the region and they are particularly diverse east of the Cascades. Elegant penstemon (*P. venustus*) and glandular penstemon (*P. glandulosus*) are also exceedingly showy drought-tolerant species that grow in eastern Washington and Oregon, as well as Idaho.

Showy penstemon is truly stunning in bloom.

Petasites frigidus • Asteraceae

Sweet coltsfoot

HABITAT/RANGE Wet places and moist forests at low to high elevations from Alaska to California and east to the Atlantic Coast. Circumboreal. Critically imperiled in Idaho and Montana. Grows on both sides of the Cascade Mountains, though mainly west.

SEASONAL INTEREST White to purplish flowers in early spring to midsummer relative to elevation. Attractive foliage.

WILDLIFE VALUE Early-season pollen and nectar source for bees, butterflies, and other pollinators. Butterfly host plant. Attracts beneficial insects. Provides nesting material for birds. Browsed by elk. Deer resistant.

CULTIVATION Shade or sun and moist to wet, humus-rich soil. Water to establish and continue to provide supplemental water as needed. Spreads by rhizomes. Can be grown in containers. Plant in rain gardens, riparian areas, bioswales, and wet spots in woodland and pollinator gardens. Mulch.

Sweet coltsfoot produces stout flowering stems up to 2 ft. tall topped with clusters of white to purplish flowers. Hungry pollinators are drawn to these early-blooming flowers. The large, attractive basal leaves unfurl after the flowers, creating a lush groundcover over moist soils. There are a few variations of this species with leaf shapes varying from rounded to triangular. *Petasites frigidus* var. *palmatus*, also referred to as *P. palmatus*, has large palmate leaves, grows mainly west of the Cascades, and is popular with gardeners. Be aware that *Tussilago farfara*, also called coltsfoot, is nonnative and invasive in the region.

Sweet coltsfoot is a lovely plant for wet places and its flowers attract butterflies and other pollinators.

Phacelia hastata • Hydrophyllaceae

Silverleaf phacelia

HABITAT/RANGE Dry, open areas from British Columbia to California and east through the Rocky Mountains. Widespread throughout the region with multiple variations occurring at low to high elevations.

SEASONAL INTEREST White to lavender flowers from late spring to late summer or early fall relative to variation and elevation.

WILDLIFE VALUE Attracts a diversity of bees, bumble bees, butterflies, and other pollinating insects. Specialist bee host. Moth host plant. Attracts beneficial insects.

CULTIVATION Sun to light shade and well-drained soil. Likes sandy or gravelly soils. Water to establish. Drought tolerant once established. Wear gloves when handling plants to avoid bristly hairs. Plant in xeric gardens, pollinator plantings, and parking strips.

Silverleaf phacelia is structurally appealing with dramatic flowers and foliage, but the true beauty of gardening with this drought-tolerant perennial is its value to pollinators. With prostrate to erect stems reaching about 20 in. and deeply veined foliage, the entire plant is covered in hairs giving it a silvery sheen. Its tightly coiled, fiddlehead-like clusters of white to lavender flowers draw clouds of pollinating insects and this species is used to improve crop pollination by supporting native bees. Many species of *Phacelia* are native to the region and make great additions to insectary plantings on organic farms. Lacy phacelia (*P. tanacetifolia*), an annual native to California and the Southwest, is widely used as a cover crop. Woodland phacelia (*P. nemoralis*) grows in shady places and is a good choice for gardens west of the Cascades.

Silverleaf phacelia is a great pollinator plant that brings structural and tonal interest to xeric gardens.

Phlox diffusa • Polemoniaceae

Spreading phlox

HABITAT/RANGE Rocky slopes and open forests in mountainous areas from British Columbia to California. Low to high elevations. Grows on both sides of the Cascade Mountains.

SEASONAL INTEREST Pink to white or blue flowers in spring to late summer relative to elevation. Semievergreen.

WILDLIFE VALUE Attracts bees, butterflies, moths, and other pollinators.

CULTIVATION Sun to light shade and very well-drained, rocky soil. Water to establish. Drought tolerant once established if properly sited. Do not plant in regularly irrigated gardens. A choice rock garden plant. Does well in containers.

Phlox is a genus of plants loved by gardeners for their ability to bloom profusely, putting on spectacular shows of color as their foliage becomes completely covered in flowers. *Phlox diffusa* does this with pink to white or blue flowers that seasonally obscure its dense, needlelike leaves. Mat-forming, woody, and growing only 4 in. tall, this semievergreen perennial makes an attractive groundcover but requires sharp drainage to thrive. There are many species of *Phlox* native to the region that would seemingly make lovely garden plants, but most tend to be finnicky in cultivation. *P. diffusa* can be more easily grown than others if given adequate drainage. Hood's phlox (*P. hoodii*) is similar and forms dense cushions of foliage and flowers in dry, open places east of the Cascades.

Phlox diffusa is a profuse bloomer that can enliven rock gardens.

Phlox longifolia • Polemoniaceae

Longleaf phlox

HABITAT/RANGE Sunny, dry areas, rocky slopes, shrub-steppe, and openings in pine forests. British Columbia to California and east to the Rocky Mountains at low to mid, occasionally high, elevations. Widespread east of the Cascade Mountains.

SEASONAL INTEREST Pink to white flowers in spring to midsummer relative to elevation.

WILDLIFE VALUE Attracts bees, butterflies, moths, and other pollinators.

CULTIVATION Sun to light shade and well-drained soil. Water only to establish. Drought tolerant once established. Do not plant in regularly irrigated gardens. May be difficult to grow west of the Cascades. Plant in rock gardens and sunny, xeric gardens.

Longleaf phlox is a plant for gardeners east of the Cascades where it can be found growing in dry places next to sagebrush and other drought-tolerant, sun-loving shrubs. As is characteristic with phlox, this perennial blooms profusely, covering itself so densely with fragrant, pink to white flowers that the foliage becomes obscured. Stems grow to over 1 ft. with linear leaves and clusters of flowers at the stem tips. Showy phlox (*Phlox speciosa*) is similar in habit and habitat but can be difficult to cultivate. Many of our native phlox are difficult to grow in moist climates like those found west of the Cascades. Gardeners there need to provide good drainage and protection from soggy conditions. Woodland phlox (*P. adsurgens*) grows west of the Cascades and provides an option for westside gardeners.

Longleaf phlox is a drought-tolerant perennial with beautiful flowers.

Polemonium californicum • Polemoniaceae

Low Jacob's ladder

HABITAT/RANGE Moist to dry forests and open places at mid to high elevations in mountainous areas from Washington to California and east to Idaho. Rare in British Columbia.

SEASONAL INTEREST Blue to purple flowers in late spring to late summer depending on location.

WILDLIFE VALUE Attracts bees, butterflies, and other pollinators. Seeds may be eaten by birds.

CULTIVATION Part shade to sun and moist to seasonally dry, well-drained soil. Water to establish and continue to provide supplemental water as needed. More drought tolerant in partial shade but can grow in sunnier sites if provided with adequate moisture. Spreads slowly by rhizomes. Can be grown in containers. Mulch lightly.

Polemoniums are popular with shade gardeners who know they bring a splash of color along with beautifully structured foliage to woodland gardens. Rhizomatous with a clumping habit, *Polemonium californicum* grows gracefully through the forest understory into open mountain areas, reaching up to 1 ft. tall, sometimes taller, with pinnate leaves and clusters of blue to purple flowers, usually with yellow centers. Showy Jacob's ladder (*P. pulcherrimum*) is similar but shorter, not rhizomatous, has a rather skunky fragrance, and is more widely distributed in the region. For gardeners with moist soils, western Jacob's ladder (*P. occidentale*) is a robust rhizomatous species that grows as much as 3 ft. tall in wet, shady sites and meadows.

Polemonium californicum has attractive foliage and flowers that attract a variety of native pollinators.

Polemonium carneum • Polemoniaceae

Great Jacob's ladder

HABITAT/RANGE Forest openings, woodlands, thickets, and streambanks from the Olympic Peninsula south to California. Grows in and west of the Cascade Mountains at low to mid elevations. This is a rare and threatened plant in California and Washington.

SEASONAL INTEREST Salmon to yellow, purple, blue, or white flowers late spring to midsummer.

WILDLIFE VALUE Attracts bees, butterflies, and other pollinators.

CULTIVATION Part shade and moist to seasonally moist, humus-rich soil. Can grow in clay soil. Water to establish and continue to provide supplemental water as needed. Supplemental summer water will keep plants green and delay dormancy. Plant on the edge of moist woodland gardens. Grows well in containers. Mulch.

Great Jacob's ladder is loved by gardeners for its appealing foliage and delicately tinted, pastel flowers. Delightfully displaying multiple shades of salmon, yellow, purple, blue, and white in its loose clusters of flowers, it is long blooming and grows 1 to 3 ft. tall with decumbent to erect stems and pinnate leaves. This shade-loving, rhizomatous herbaceous perennial is easy to grow in relatively moist, partly shaded sites. *Polemonium carneum* is rare in parts of its range and wild populations are under threat; always make sure plants are ethically propagated and never taken from the wild.

Polemonium carneum is an attractive shade-loving perennial with showy flowers in a range of pastel colors.

Potentilla gracilis • Rosaceae

Slender cinquefoil

HABITAT/RANGE Widespread and highly variable species that occurs in many different habitats. Alaska to California and east beyond the Rocky Mountains at low to high elevations. Grows on both sides of the Cascade Mountains.

SEASONAL INTEREST Yellow flowers in summer.

WILDLIFE VALUE Attracts bees, butterflies, and other pollinators. Attracts beneficial insects. Host plant for multiple species of butterflies and moths. Specialist bee host. Seeds eaten by birds.

CULTIVATION Sun to light shade and moist to seasonally moist, well-drained soil. Grows in a variety of soil types. Water to establish and continue to provide supplemental water as needed. Tolerates seasonally dry conditions. Spreads by seed; cut back before seed ripens to curtail spreading if needed. Plant in wildflower meadows, pollinator plantings, and rain gardens. Can be grown in containers. Mulch.

Slender cinquefoil is an important plant for pollinators and a perfect choice for wildflower meadows. This herbaceous perennial is highly variable across its range, but generally grows around 2 ft. tall with mostly basal foliage and clusters of showy yellow flowers on branching stalks. The leaves are palmate, toothed, and often woolly on the undersides. Do not confuse with sulphur cinquefoil (*Potentilla recta*), which is an invasive species from Eurasia that looks similar. There are many species of *Potentilla* native to the region that make great pollinator plants.

Slender cinquefoil is a colorful wildflower and a great plant for pollinators.

Prosartes hookeri • Liliaceae

Hooker's fairybells

HABITAT/RANGE Moist, shady areas and forests from British Columbia to California and east to Montana. Low to mid elevations. Grows on both sides of the Cascade Mountains.

SEASONAL INTEREST White flowers in spring to summer relative to elevation. Red to orange-red berries.

WILDLIFE VALUE Attracts bees and other pollinators. Fruits eaten by birds and small mammals.

CULTIVATION Shade to part sun and moist, humus-rich soil. Water to establish and continue to provide supplemental water as needed. Tolerates seasonally dry conditions once established if properly sited. Rhizomatous but spreads slowly. Mulch.

Hooker's fairybells is a lush and lovely herbaceous perennial for moist, shady woodland gardens. This understory plant grows 2 to 3 ft. tall with branching stems and clasping, strongly veined leaves. Pendulous, bell-shaped, white flowers bloom in pairs at the stem tips and develop into attractive bright red to orange-red berries. Formerly classified as *Disporum hookeri*, this is a variable species across its range. Fairy lantern (*Prosartes smithii*), which grows entirely west of the Cascades, is similar with larger, unflared flowers and yellow to orange or red berries. Clasping twisted-stalk (*Streptopus amplexifolius*) is a close relative that grows in the same habitats and looks similar to Hooker's fairybells, but has flowers and yellow to red berries that hang down from its leaf axils. Rosy twisted-stalk (*S. lanceolatus*) grows at higher elevations and is shorter with pink to yellowish flowers and red berries.

Hooker's fairybells is a shade-loving perennial with pendulous white flowers and brightly colored fruits.

Prunella vulgaris var. *lanceolata* • Lamiaceae

Self-heal

HABITAT/RANGE Moist places and disturbed sites at low to mid elevations. Widespread throughout North America, Europe, Asia, and other parts of the world. Both native and introduced variations occur in the region.

SEASONAL INTEREST Blue-violet flowers late spring to early fall.

WILDLIFE VALUE Attracts bees, bumble bees, butterflies, hummingbirds, and other pollinators. Attracts beneficial insects. Seeds eaten by birds.

CULTIVATION Sun to shade and moist soil. Grows in a variety of conditions but prefers some moisture. Water to establish and continue to provide supplemental water as needed. Spreads aggressively by seed and short rhizomes. Deadhead or cut back to encourage new blooms. Remove seed heads before ripe to curtail spreading if needed. Can be used as a lawn alternative. Plant along pond margins, in insectary plantings, as a groundcover in orchards, or places where covering soils and keeping out weeds is desired. Grows well in containers.

Prunella vulgaris is lovely but spreads aggressively in irrigated gardens. Nonetheless, it can be the perfect plant in the right situation and creates a pollinator-friendly groundcover. Typically growing about 1 ft. tall, this herbaceous perennial has erect stems and spikes of blue-violet flowers. In moist soils plants bloom perpetually through summer if periodically cut back. Be aware that an introduced Eurasian variation (*P. vulgaris* var. *vulgaris*) grows in the region and can be difficult to distinguish from the native variation (*P. vulgaris* var. *lanceolata*).

Prunella vulgaris is a pollinator-friendly groundcover with attractive spikes of blue-violet flowers.

Ranunculus occidentalis • Ranunculaceae

Western buttercup

HABITAT/RANGE Meadows, grasslands, vernal pools, and open woodlands from Alaska to California. Low to mid elevations. Grows on both sides of the Cascade Mountains, though mainly west.

SEASONAL INTEREST Yellow flowers in early spring to midsummer relative to elevation.

WILDLIFE VALUE Attracts bees, butterflies, and other pollinators. Specialist bee host. Attracts beneficial insects. Seeds eaten by birds. Deer resistant.

CULTIVATION Sun to part shade and moist to seasonally dry soil. Grows in a variety of soil types including clay. Water to establish. Drought tolerant once established if properly sited. Goes dormant after setting seed. Can be a short-lived perennial but reseeds itself readily. May be easiest to establish by seed. Plant in vernally moist areas, wildflower meadows, oak woodlands, and insectary plantings.

This herbaceous perennial is stunning when blooming en masse in vernally moist meadows. The sight of it can make you feel like a kid again as you watch pollinators dart about gathering nectar. Spreading buttercups in open areas can bring beauty and biodiversity to your neighborhood. Western buttercup grows about 1 ft. tall with deeply lobed basal leaves and multiple shiny yellow flowers. There are many buttercups in the Pacific Northwest that range from rare and endemic to widely distributed native species, as well as some aggressive introduced species.

Meadows and hillsides emit their own sunshine when western buttercup is in bloom.

Rudbeckia occidentalis • Asteraceae

Western coneflower

HABITAT/RANGE Riparian areas, wetlands, seeps, moist meadows, and forest openings. Washington to California and east to Montana at moderately low to high elevations. Grows mainly east of the Cascade Mountains.

SEASONAL INTEREST Flowers in summer. Seed heads remain structurally appealing.

WILDLIFE VALUE Attracts bees, butterflies, and other pollinators. Attracts beneficial insects. Seeds eaten by birds. Provides cover.

CULTIVATION Sun to part shade and moist to wet, humus-rich soil. Plant in places with consistent moisture like pond margins, wetlands, streambanks, and irrigated gardens. Can be grown in containers. Mulch.

Western coneflower is a curious composite flower in that it lacks any ray florets (what look like petals), having only a prominent, elongated cone of darkly colored disc florets subtended by a ring of green bracts. This may not make for a very colorful flower, but it certainly makes for an interesting one and even a good cut flower. These seemingly petalless flowers grow on long stems that reach as much as 5 to 6 ft. tall in ideal conditions. The large leaves are attractive and provide cover for wildlife while the flowers and seeds feed pollinators and birds. This is a popular and robust herbaceous perennial with a few cultivars available. Choose plants wisely and favor planting true natives.

Western coneflower is a robust riparian plant and a rayless cousin of black-eyed Susan that attracts beneficial insects like these braconid wasps.

Sagittaria latifolia • Alismataceae

Wapato

HABITAT/RANGE Edges and shallows of ponds and lakes, wetlands, and ditches. Vancouver Island to central California and scattered locations east to Idaho. Occurs primarily east of the Rocky Mountains. Grows on both sides of the Cascade Mountains, though mostly west, at low elevations.

SEASONAL INTEREST White flowers in summer to early fall.

WILDLIFE VALUE Attracts a variety of pollinators and beneficial insects. Seeds eaten by birds. An important food for wildlife including waterfowl. Provides cover. Browsed by deer.

CULTIVATION Sun to light shade and shallow standing water or saturated soil. Prefers loam or clay soils; does not like sandy soils. Spreads by rhizomes and seed. Plant along ponds and in wetlands. Can be grown in containers kept in trays of water. Be sure starts are ethically propagated and not taken from the wild. Protect plants from predation until established.

Wapato is an important wetland plant to the people and wildlife of the Pacific Northwest. With invasive species outcompeting native plants in wetland habitats that are already in general decline, this plant needs our help. The many benefits of planting wapato include its ability to improve soil and water health. This attractive herbaceous perennial grows 1 to 3 ft. tall with whorls of white flowers on long stems and large arrow-shaped leaves. It creates colorful tubers in pastel shades and is a favored food of waterfowl. Arumleaf arrowhead (*Sagittaria cuneata*), also called wapato, grows mostly east of the Cascades.

Wapato is a beautiful wetland plant with large, arrow-shaped leaves.

Saxifraga mertensiana • Saxifragaceae

Woodland saxifrage

HABITAT/RANGE Wet cliffs, seeps in rock faces, and gravelly streambanks from Alaska south to California and east to Idaho and Montana. Low to high elevations. Grows on both sides of the Cascade Mountains.

SEASONAL INTEREST White flowers in early spring to summer relative to elevation.

WILDLIFE VALUE Early-season forage for insect pollinators.

CULTIVATION Part shade to sun and rocky, moist but well-drained soil. Water to establish. Tolerates seasonally dry conditions once established if properly sited. Providing supplemental water may keep plants from going dormant in the heat of summer. Clumps increase slowly by short rhizomes. Can be grown in containers.

Woodland saxifrage is a lovely herbaceous perennial for moist, partly shaded rock gardens and walls. It forms attractive clumps of succulent basal leaves that are rounded and toothed. Branched flowering stalks grow up to 16 in. tall with sprays of white flowers whose pink anthers make the flowers even more pleasing. Sometimes the flowers are replaced by small pinkish bulbils (bulblike structures) that fall off and become new plants. *Saxifraga* is a genus loved by rock and alpine gardeners and there are many species native to the region, though some are difficult to cultivate.

Woodland saxifrage produces attractive mounds of foliage and sprays of white flowers.

Sedum oreganum • Crassulaceae

Oregon stonecrop

HABITAT/RANGE Rocky sites and mountainous and coastal areas from Alaska to California. Low to mid elevations. Grows mainly in and west of the Cascade Mountains.

SEASONAL INTEREST Yellow flowers in summer. Evergreen.

WILDLIFE VALUE Attracts bees, butterflies, and other pollinators. Attracts beneficial insects. A few butterflies specialize on sedums as larval hosts. Seeds and succulent foliage eaten by birds and other wildlife.

CULTIVATION Sun to part shade and well-drained, moist to dry soil. Prefers rocky sites and lightly amended, gritty soil. Water to establish. Drought tolerant once established if properly sited but will benefit from occasional summer water, especially in arid areas. Vegetative growth roots readily and can be used to start new garden plants. A choice plant for rock gardens and rock walls. Good for green roofs and planters. Responds well to a light gravel mulch.

Sedums are known as the ultimate rock garden plant, but what some gardeners may not realize is that native sedums are also excellent pollinator plants. *Sedum oreganum* is a succulent evergreen groundcover that grows in rocky, well-drained sites where its showy yellow flowers attract many pollinators. Its flowering stems grow to around 6 in. tall over glossy, green to reddish rosettes of foliage. Spreading stonecrop (*S. divergens*) is another easy to grow and attractive native sedum that brings interesting texture and color to the edges of butterfly gardens and rock walls.

Sedum oreganum loves to grow in rocky places.

Sedum spathulifolium • Crassulaceae

Broadleaf stonecrop

HABITAT/RANGE Rocky sites and cliffs from British Columbia to California. Low to mid elevations. Grows mainly in and west of the Cascade Mountains.

SEASONAL INTEREST Yellow flowers in spring to summer relative to elevation. Evergreen.

WILDLIFE VALUE Attracts bees, butterflies, and other pollinators. Attracts beneficial insects. A few butterflies specialize on sedums as larval hosts. Seeds and succulent foliage eaten by birds and other wildlife.

CULTIVATION Part shade to sun and well-drained, moist to dry soil. Prefers rocky sites and gritty soil. Water to establish. Drought tolerant once established if properly sited. A choice rock garden plant. Vegetative growth roots readily; tuck into soil between the rocks of a rock wall. Can be grown in containers. Light gravel mulch.

Broadleaf stonecrop is a particularly striking sedum that puts on a serious show of color just in its succulent foliage. It forms loose to dense mats of thick, silvery bluish gray to red or green leaves in basal rosettes. Flowering stems grow as much as 8 in. tall and may be vividly colored themselves, sometimes even neon pink, with multiple bright yellow, starlike blooms at their tips. *Sedum spathulifolium* is an evergreen groundcover, but the foliage may diminish slightly during the heat of summer. It is an attractive plant for rock gardens, butterfly gardens, and rock walls. There are many cultivars and selections of this popular and variable perennial commonly available.

Broadleaf stonecrop has colorful foliage and bright yellow, starlike flowers.

Sidalcea oregana • Malvaceae

Oregon checkermallow

HABITAT/RANGE Meadows, riparian areas, shrub-steppe, pine forests, and oak woodlands. Washington to California and east to Wyoming at low to high elevations. Grows mainly east of the Cascade Mountains.

SEASONAL INTEREST Pink flowers late spring to summer.

WILDLIFE VALUE Important butterfly and pollinator plant. Attracts hummingbirds. Butterfly host plant. Specialist bee host. Attracts beneficial insects. Seeds may be eaten by birds. Browsed by deer.

CULTIVATION Sun to part shade and well-drained, humus-rich soil. Provide some shade in arid areas. Water to establish and continue to provide supplemental water as needed. Tolerates seasonally dry conditions but prefers moderate moisture. Great for moist meadows and butterfly gardens. Mulch.

Oregon checkermallow is a robust perennial that grows 2 to 5 ft. tall with showy racemes of pink, hibiscus-like flowers and palmately lobed leaves. Checkermallows are important butterfly plants that resemble hollyhocks and add a cottage garden look to the landscape. Western North America is home to the world's checkermallows, and while they may grow easily in gardens, many species are rare or becoming rare and endangered due to loss of habitat. It is important gardeners plant and proliferate these species in areas where they were once abundant, but it is vital plants are ethically sourced. *Sidalcea oregana* is the best choice for gardens east of the Cascades. Coastal gardeners will want to provide a refuge for Henderson's checkermallow (*S. hendersonii*), an imperiled, but commercially available, plant. Meadow checkermallow (*S. campestris*) is a good choice for the Willamette Valley.

Oregon checkermallow grows robustly in moist meadows and attracts butterflies and other pollinators.

Sisyrinchium idahoense • Iridaceae

Idaho blue-eyed grass

HABITAT/RANGE Moist sites, meadows, prairies, riparian areas, wetland edges, and open woodlands. Southern British Columbia to California and east to the Rocky Mountains at low to fairly high elevations. Widespread. Grows on both sides of the Cascade Mountains.

SEASONAL INTEREST Blue-violet flowers in midspring to midsummer relative to elevation.

WILDLIFE VALUE Attracts bees, butterflies, and other pollinators. Attracts beneficial insects. Seeds eaten by birds.

CULTIVATION Sun to part shade and wet to seasonally moist soil. Water to establish and continue to provide supplemental water as needed. Tolerates seasonally dry conditions. Grows easily from seed but seedlings may not compete well with other vegetation. Mature garden plants can be divided. Plant in moist meadows, irrigated perennial beds, riparian areas, and rain gardens. Can be grown in containers. Mulch lightly.

Idaho blue-eyed grass is a small perennial with flattened, grasslike leaves and pale to dark blue-violet flowers. Plants grow 6 to 16 in. tall in attractive tufts that slowly increase over time. This herbaceous perennial is a variable and adaptable species that prefers moist soils but can also tolerate seasonally dry conditions. Western blue-eyed grass (*Sisyrinchium bellum*) is similar but occurs mostly west of the Cascades in Oregon and south through California. Golden-eyed grass (*S. californicum*) is a coastal species with yellow flowers that can spread vigorously by seed.

Idaho blue-eyed grass is a sweet little perennial with flowers that attract small bees.

Solidago spp. • Asteraceae

Goldenrod

HABITAT/RANGE Open areas, meadows, riparian areas, thickets, open woods, and disturbed sites at low to high elevations. Widespread.

SEASONAL INTEREST Yellow flowers summer to fall.

WILDLIFE VALUE Important late-season nectar and pollen source for many pollinators. Attracts beneficial insects. Specialist bee host. Butterfly host plant. Old stalks used by cavity-nesting bees. Seeds eaten by birds. Provides cover. Browsed by deer and other wildlife.

CULTIVATION Sun to part shade and moderately moist soil. Grows in a variety of soil types. Water to establish and continue to provide supplemental water as needed. Some species spread vigorously by rhizomes or seed. Drier soils restrict rhizomatous growth. Cut back seed heads before ripe to curtail spreading if needed but leave some seed for the birds and 8 to 24 in. of old flower stalks for cavity-nesting bees. Great for pollinator and insectary plantings. Useful for holding soils. Grows well in planters but dominates. Mulch.

Goldenrods are colorful, late-blooming, easy-to-grow herbaceous perennials that bring beauty and biodiversity to the landscape. Useful as cut flowers, dye plants, and even tea herbs, these important pollinator plants are powerhouses for attracting beneficial insects and deserve a place in farms and gardens. Western Canada goldenrod (*Solidago lepida*) and West Coast goldenrod (*S. elongata*) are common and attractive species with showy clusters of yellow flowers that gardeners may enjoy. Contrary to popular belief, goldenrod does not commonly cause allergic reactions to its pollen. This is a taxonomically complex genus whose variation complicates identification.

West Coast goldenrod (*Solidago elongata*) provides abundant late-season color, as well as forage for pollinators and birds.

Stachys cooleyae • Lamiaceae

Cooley's hedgenettle

HABITAT/RANGE Moist places, riparian areas, wetlands, forests, and roadsides from British Columbia to southern Oregon at low to mid elevations. Grows mainly from the east base of the Cascade Mountains to the Pacific Coast.

SEASONAL INTEREST Bright pink to purplish flowers in summer.

WILDLIFE VALUE Attracts bees, bumble bees, butterflies, hummingbirds, and other pollinators. Moth host plant. Stabilizes riparian soils, which helps to create healthy habitat for aquatic life. Browsed by deer.

CULTIVATION Part shade to sun and moist to wet soil. Water to establish and continue to provide supplemental water as needed. Tolerates occasional flooding. Spreads by rhizomes. Great for wet meadows, riparian areas, rain gardens, bioswales, forest edges, irrigated beds, and hummingbird gardens. Mulch.

Stachys cooleyae is a beautiful plant for moist places. The bright pink to purplish, tubular flowers have an orchid-like appearance and are a magnet for hummingbirds and other pollinators. They bloom in whorls at the tips of the stems bringing long-lasting summer color to streambanks and lakeshores. Another lovely aspect of this herbaceous perennial is its pleasantly aromatic foliage. Cooley's hedgenettle can grow as much as 5 ft. tall in ideal conditions. Also classified as *S. chamissonis* var. *cooleyae*.

Stachys cooleyae is a beautiful, aromatic perennial with tubular flowers that attract hummingbirds.

Symphyotrichum foliaceum • Asteraceae

Leafybract aster

HABITAT/RANGE Moist meadows, forest openings, and riparian areas from British Columbia to California and east to the Rocky Mountains. Generally found at mid to high elevations. Grows primarily in and east of the Cascade Mountains in Oregon, on both sides in Washington.

SEASONAL INTEREST Blue-violet to rose-purple flowers with yellow centers midsummer to fall, earlier at lower elevations.

WILDLIFE VALUE Late-season nectar and pollen source for bees, butterflies, and other pollinators. Butterfly host plant. Many solitary bees specialize on plants in the family Asteraceae. Attracts beneficial insects. Seeds eaten by birds. Browsed by deer.

CULTIVATION Sun to part shade and moist, humus-rich soil. Grows in many soil types. Water to establish and continue to provide supplemental water as needed. Prefers some shade in drier sites. Rhizomatous. Deadhead. Plant in sunny, irrigated perennial beds, pollinator gardens, rain gardens, and moist meadows. Can be grown in containers. Mulch.

Asters are important plants for pollinators. Blooming late in the growing season they provide nectar and pollen that fuels the migration of monarch butterflies and fills the winter larder of native bees. The seeds are a favorite food of birds and plants attract beneficial insects. Leafybract aster is a lovely and variable species that grows 1 to 2 ft. tall with erect to decumbent leafy stems. It has blue-violet to rose-purple composite flowers with yellow centers that can bloom into fall. Previously classified as *Aster foliaceus*.

Native asters are beautiful late-blooming plants perfect for pollinator and butterfly gardens.

Synthyris reniformis/Veronica regina-nivalis • Plantaginaceae

Snow queen

HABITAT/RANGE Moist forests from Washington to California. Low to mid elevations. Grows west of the crest of the Cascade Mountains.

SEASONAL INTEREST Blue to purplish flowers early to late spring. Typically evergreen.

WILDLIFE VALUE Very early-season nectar and pollen source for pollinators. Attracts beneficial insects.

CULTIVATION Full to part shade and well-drained, moist to vernally moist, humus-rich soil. Water to establish and continue to provide supplemental water as needed. Drought tolerant once established if properly sited in moist forests west of the Cascades. Mulch lightly but do not bury this low-growing plant.

Snow queen is one of the first plants to bloom in westside forests, gaining its common name by flowering as snow melts in early spring. It is a sweet little understory plant that grows up to 6 in. tall with blue to purple, bell-shaped flowers over basal rosettes of heart-shaped leaves. This woodland perennial is typically evergreen and provides an early-season nectar and pollen source for pollinators. Mountain kittentails (*Synthyris missurica/Veronica missurica*) is also an early herald of spring with similar but showier and taller flowers. It grows in moist forests and slopes east of the Cascades and is most prolific around northern Idaho. Molecular studies show the genus *Synthyris* should be included in the genus *Veronica* and both names are used here.

Snow queen is an early-blooming wildflower found in forests west of the Cascade Mountains.

Tellima grandiflora • Saxifragaceae

Fringecup

HABITAT/RANGE Moist to seasonally moist, shady sites, westside forests, and streambanks. Alaska to California and east to Idaho at low to mid elevations. Grows mainly west of the crest of the Cascade Mountains.

SEASONAL INTEREST White to pink flowers in spring to summer relative to elevation. Evergreen in mild winters.

WILDLIFE VALUE Attracts bees, hummingbirds, and other pollinators. Attracts beneficial insects. Deer resistant.

CULTIVATION Part to full shade and moist to seasonally moist, humus-rich soil. Water to establish. Drought tolerant once established if properly sited. Can grow under conifers. Spreads by seed. Cut back stalks before seed ripens to curtail spreading if needed. Prone to powdery mildew; if affected cut back to encourage a fresh flush of fall foliage. Plant in woodland gardens and shady perennial beds. Grows well in containers. Mulch.

Fringecup is a perfect plant for shady gardens. What is not to love about showy 2 ft. tall spikes of feathery flowers that start out white and slowly turn pale to deep pink. Or attractive, mounding basal foliage that remains somewhat evergreen in mild winters and bronzes nicely in cold weather. This herbaceous perennial has a lot going for it. In some sites, particularly drier ones, it may remain clumped, while in moist understories it spreads more vigorously. Quite eye-catching when blooming en masse, it provides ample forage for pollinators and garden plants make a nice cut flower.

Fringecup is an easy-to-grow, shade-loving perennial with spikes of white flowers that fade to pink.

Thalictrum occidentale • Ranunculaceae

Western meadowrue

HABITAT/RANGE Moist forests and meadows and streambanks. Southeast Alaska to Northern California and east to Montana at low to high elevations. Grows on both sides of the Cascade Mountains.

SEASONAL INTEREST Blooms spring to summer relative to elevation. Lacy foliage.

WILDLIFE VALUE Wind pollinated. Some insects collect pollen.

CULTIVATION Full to part shade and moist, humus-rich soil. Water to establish and continue to provide supplemental water as needed. Plant in moist woodland gardens or shady riparian areas. Mulch.

While western meadowrue may have subtle flowers, its exquisitely beautiful foliage looks lovely in moist, shaded gardens. This herbaceous perennial is dioecious, meaning plants are either male or female. Male plants tend to be showier due to their many clusters of pendulous stamens, dangling so the wind can easily catch and carry their pollen to surrounding female plants, which have clusters of purplish pistils. The erect, branching stems grow up to 3 ft. tall with delicate, lacy foliage. There are other species of *Thalictrum* found in the region, *T. occidentale* being the most widespread and common in Oregon and Washington.

Western meadowrue brings a layer of lacy foliage to the understory of moist woodland gardens.

Tolmiea menziesii • Saxifragaceae

Piggyback plant

HABITAT/RANGE Moist woodlands and shaded riparian areas from Alaska to Oregon. Low to mid elevations. Grows west of the crest of the Cascade Mountains.

SEASONAL INTEREST Greenish purple to maroon flowers in spring to summer relative to elevation. Evergreen in mild winters.

WILDLIFE VALUE Attracts bees, butterflies, and other pollinators. Attracts beneficial insects. Deer resistant.

CULTIVATION Full to part shade and moist but well-drained, humus-rich soil. Water to establish and continue to provide supplemental water as needed. Tolerates seasonally dry conditions once established if properly sited. Spreads by rhizomes. Use as a groundcover in moist woodland gardens. Can be grown in containers and as a houseplant. Mulch.

Piggyback plant, also called youth-on-age, carpets the floor of westside forests with distinctly veined leaves and loose racemes of greenish purple to maroon flowers that grow 1 ft. tall, sometimes taller. Blooming in patches, the flowers are quite eye-catching, and though they are small, with a closer look you will see they are as intricate and flamboyant as an orchid. Not only does this herbaceous perennial spread by rhizomes, it also creates new plantlets on top of older leaves that root when they touch the ground. This distinctive way of propagating itself has inspired its common names.

Tolmiea menziesii is a lovely groundcover for moist, shady woodland gardens west of the Cascades.

Trifolium macrocephalum • Fabaceae

Big-head clover

HABITAT/RANGE Rocky, dry sites, grasslands, shrub-steppe, and open pine woodlands. Washington to California and east to Idaho at low to mid elevations. Grows east of the Cascade Mountains.

SEASONAL INTEREST Pale to deep pink and white flowers spring to early summer.

WILDLIFE VALUE Attracts bees, butterflies, and other pollinators. Attracts beneficial insects. Butterfly host plant. Specialist bee host. Seeds eaten by birds. Browsed by deer and other wildlife.

CULTIVATION Sun to light shade and well-drained, vernally moist soil. Water to establish. Drought tolerant once established. Goes dormant after setting seed. Rhizomatous but slow to spread. Plant in sunny, xeric pollinator gardens, rock gardens, and dry wildflower meadows. Best grown east of the Cascades.

Trifolium macrocephalum is one of the showiest native clovers in the region. This drought-tolerant herbaceous perennial grows around 6 in. tall with large clusters of pink and white, pealike flowers and palmately compound leaves with a light stripe across each leaflet. While it is a great plant for gardens east of the Cascades, it is not widely available in nurseries. There are, however, many native species of *Trifolium*, some more commonly available such as tomcat clover (*T. willdenovii*) and springbank clover (*T. wormskioldii*), the latter of which is great for moist sites. Clovers fix nitrogen into the soil making them useful as pollinator-friendly cover crops. Many introduced species have escaped cultivation and naturalized while populations of native species are in decline and should be planted more.

True to its name, the flowers of big-head clover are unusually large.

Trillium ovatum • Melanthiaceae

Western trillium

HABITAT/RANGE Moist forests and shady streambanks from British Columbia to California and east to Montana. Low to mid elevations. Grows on both sides of the Cascade Mountains.

SEASONAL INTEREST White flowers that fade to pink, early spring to summer relative to elevation.

WILDLIFE VALUE Attracts bees, moths, and other pollinators. Attracts beneficial insects. Seeds have a fatty protein package that attracts ants who disperse the seed. Fruits eaten by birds and mammals.

CULTIVATION Full to part shade and moist to seasonally moist, humus-rich soil. Prefers slightly acidic soil. Water to establish. Tolerates seasonally dry conditions once established if properly sited. Goes dormant in summer. Spreads slowly by seed and short rhizomes. Takes many years to flower from seed. Do not use as a cut flower as this will impede the plant's ability to bloom the following year or kill it entirely. Protect from slugs. A woodland plant that likes a leaf litter mulch.

Western trillium is a harbinger of spring in the Pacific Northwest. This is a popular herbaceous perennial for moist, shady woodland gardens that grows around 1 ft. tall with a single, showy, white flower and whorl of three large leaflike bracts. Other species popular with gardeners are giant trillium (*Trillium albidum*), which has mottled "leaves," and giant purple trillium (*T. kurabayashii*), which has dark red flowers. These species may be easily found in nurseries but are rare in all or parts of their range. Make sure plants are ethically sourced and never poached from the wild.

Western trillium is an iconic spring wildflower of western forests.

Triteleia grandiflora • Asparagaceae

Large-flowered triteleia

HABITAT/RANGE Coastal bluffs, prairies, grasslands, oak savannas, shrub-steppe, and open woodlands. Southern British Columbia to California and east to Montana at low to high elevations. Grows on both sides of the Cascade Mountains. Listed as rare in California.

SEASONAL INTEREST Blue to white or purple flowers in spring to summer relative to elevation.

WILDLIFE VALUE Attracts bees, butterflies, and other pollinators. Bulbs eaten by mammals.

CULTIVATION Sun to light shade and well-drained, vernally moist soil. Goes dormant after setting seed. Water to establish but only sparingly after it goes dormant. Prefers dry summer conditions and saturated soils can cause bulbs to rot. Drought tolerant once established. Plant in wildflower meadows, rock gardens, and xeric perennial beds. Mulch lightly.

Triteleia grandiflora is a variable and very beautiful perennial bulb whose clusters of large tubular flowers range in color from purple to blue, white, or bicolored. The flowers bloom on top of long leafless stems that grow to around 2 ft. tall with grasslike basal leaves. This ephemeral wildflower looks lovely in oak woodlands or sunny meadows. Plant it among bunchgrasses and drought-tolerant perennials. Since this plant is variable across its range, gardeners should seek to find plants ethically propagated from local or regionally appropriate stock. Rare in some places, the ethical propagation of this plant is emphasized.

Triteleia grandiflora is an enchanting drought-tolerant wildflower.

Triteleia hyacinthina • Asparagaceae

Hyacinth brodiaea

HABITAT/RANGE Vernally moist grasslands and meadows from southwestern British Columbia to California and east to Idaho. Low to mid elevations. Grows on both sides of the Cascade Mountains.

SEASONAL INTEREST White to pale blue flowers in late spring to summer.

WILDLIFE VALUE Attracts bees, butterflies, and other pollinators. Bulbs eaten by mammals.

CULTIVATION Sun to light shade and vernally moist soil. Goes dormant after setting seed. Water to establish, but sparingly after it goes dormant. Drought tolerant once established. Easy to cultivate. Increases by seed and bulb offsets and may naturalize in ideal conditions. Plant in wildflower meadows and sunny gardens. Mulch lightly.

Patches of *Triteleia hyacinthina* can light up vernally moist meadows, though sadly many of the meadows where it once flourished have been lost to development and agriculture. This lovely perennial bulb has grasslike leaves and a sturdy stem that grows 1 to 2 ft. tall with terminal clusters of showy, white to pale blue flowers, each with a bluish green midvein. Also called fool's onion, it resembles an onion flower but lacks the fragrance. This is a great plant for sunny, seasonally moist sites on either side of the Cascades. Make sure plants are ethically sourced.

Hyacinth brodiaea is a beautiful wildflower that is easily cultivated.

Valeriana sitchensis • Valerianaceae

Sitka valerian

HABITAT/RANGE Moist places, meadows, and forest openings in mountainous areas from Alaska to Northern California and east to Montana. Mostly at mid to high elevations. Grows on both sides of the Cascade Mountains.

SEASONAL INTEREST White to pale pink flowers in spring to late summer relative to elevation.

WILDLIFE VALUE Attracts bees, butterflies, and other pollinators. Browsed by deer, elk, and other wildlife.

CULTIVATION Sun to part shade and moist soil. Water to establish and continue to provide supplemental water as needed. A mountain-dwelling species that prefers cool, moist soils. Provide some shade in hot areas. Rhizomatous. Plant along woodland edges and in irrigated gardens. Can be grown in containers. Mulch.

Sitka valerian fills mountain meadows with its attractive foliage and fragrant flowers that smell like heliotrope, which is why some call it mountain heliotrope. Its white to pale pink flowers have long, extended stamens and bloom in clusters on stems that grow 1 to 3 ft. tall. Although this herbaceous perennial is typically found at higher elevations, it is easily grown lower down if given a cool spot and adequate moisture. A medicinal plant that should be grown to be harvested, this beautiful perennial should be incorporated into farms and gardens more than it is. Scouler's valerian (*Valeriana scouleri*, still classified by some as *V. sitchensis* var. *scouleri*) is similar and grows mainly west of the Cascades at low to mid elevations.

Sitka valerian has clusters of white to pale pink flowers with a lovely fragrance similar to heliotrope.

Vancouveria hexandra • Berberidaceae

Inside-out flower

HABITAT/RANGE Forests and forest edges from the Puget Trough to Northern California. Grows west of the Cascade Mountains at low to mid elevations.

SEASONAL INTEREST White flowers in spring to summer relative to elevation. Attractive foliage.

WILDLIFE VALUE Attracts bees and other pollinators. Attracts beneficial insects. Seeds eaten by birds. Deer resistant.

CULTIVATION Full to part shade and moist to seasonally dry, humus-rich soil. Water to establish. Grows in seasonally dry soils under conifers or along forest edges but benefits from some supplemental water in summer. Spreads unaggressively by rhizomes. Use as a groundcover or border plant in shady woodland gardens. Mulch.

Inside-out flower has white, nodding, parachute-like flowers on stems that grow to about 16 in. tall. While these small flowers are quite fetching, many consider the most appealing quality of this tough but seemingly delicate plant to be its foliage. The compound leaves have trilobed leaflets that resemble the feet of a duck. Appropriately, duck's foot is another common name for the plant. This herbaceous perennial thrives in the understory of forests west of the Cascades where its leaves create an elegant groundcover in light to dark shades of green.

Inside-out flower is a shade-loving groundcover with attractive foliage.

Viola glabella • Violaceae

Stream violet

HABITAT/RANGE Moist woodlands and streambanks from Alaska to California and east to Montana. Low to high elevations. Widespread. Grows on both sides of the Cascade Mountains.

SEASONAL INTEREST Yellow flowers in early spring to summer relative to elevation.

WILDLIFE VALUE Early-season nectar source. Attracts bees, butterflies, and other pollinators. Attracts beneficial insects. Host plant for multiple species of butterflies. Browsed by deer.

CULTIVATION Full to part shade and moist to wet, humus-rich soil. Water to establish and continue to provide supplemental water as needed. Spreads vigorously by seed and rhizomes to form dense colonies in ideal conditions. Use in shaded woodland gardens, riparian areas, rain gardens, and butterfly gardens. Can be grown in containers. Mulch.

Stream violet is a lovely low-growing groundcover for moist, shady gardens. Spreading by rhizomes and explosively released seed, this herbaceous perennial grows up to 1 ft. tall with heart-shaped leaves and yellow flowers with five petals, the lower three marked with dark nectar lines leading pollinators to their prize. The leaves and flowers are edible but only in small quantities. There are many species of *Viola* native to the Pacific Northwest. Early blue violet (*V. adunca*) tolerates a wider range of conditions and is the larval host for the threatened Oregon silverspot butterfly. Sagebrush violet (*V. trinervata*) is a beautiful choice for gardeners east of the Cascades. Evergreen violet (*V. sempervirens*) has leaves that persist throughout the winter.

Viola glabella's cheerful yellow flowers brighten woodlands in early spring.

Wyethia amplexicaulis • Asteraceae

Northern mule's ears

HABITAT/RANGE Seasonally moist areas in shrub-steppe, meadows, and open pine woodlands. Northern Washington south through Oregon to Nevada and east to the Rocky Mountains at moderately low to fairly high elevations. Grows east of the Cascade Mountains.

SEASONAL INTEREST Yellow flowers in late spring to midsummer depending on location.

WILDLIFE VALUE Attracts bees, butterflies, and other pollinators. Many solitary bees specialize on plants in the family Asteraceae. Attracts beneficial insects. Seeds eaten by birds. Provides cover. Browsed by deer and other wildlife.

CULTIVATION Sun to light shade and vernally moist soil. Grows well in clay soils, but also grows in well-drained, loamy soils. Water to establish. Drought tolerant once established. Plant in sunny pollinator gardens, wildflower meadows, and seasonally moist sites east of the Cascades. Mulch lightly.

This robust herbaceous perennial is closely related and looks similar to balsamroot. Unlike balsamroot, which has predominantly basal leaves, northern mule's ears has both basal and stem leaves. The large leaves are smooth and shiny with pronounced veins. Its sunflower-like yellow flowers attract a variety of pollinators. This beauty grows to about 2.5 ft. tall from a stout taproot. It is an excellent choice for wildlife- and pollinator-friendly gardens east of the Cascades. Gardeners west of the Cascades will want to plant narrowleaf mule's ears (*Wyethia angustifolia*). White-head mule's ears (*W. helianthoides*) grows in eastern Oregon and central Idaho and has lovely white flowers.

Northern mule's ears has showy sunflower-like blooms.

Grasses and Grasslike Plants

Achnatherum hymenoides/Eriocoma hymenoides • Poaceae

Indian ricegrass

HABITAT/RANGE Grasslands, arid plains, shrub-steppe, and sunny, well-drained, often sandy or rocky sites from the Yukon to Mexico and east to the Great Plains. Low to mid elevations. Widespread east of the Cascade Mountains.

SEASONAL INTEREST Flowers late spring to early summer. Attractive seed heads.

WILDLIFE VALUE Seeds eaten by many species of birds and small mammals. Larval host for skipper butterflies, many of which specialize on plants in the family Poaceae. Provides cover and nesting materials. Browsed by elk, deer, rabbit, and small mammals.

CULTIVATION Sun to light shade and dry, rocky or sandy soil with good drainage. Water to establish. Drought tolerant once established. A short-lived perennial; save seed from garden plants to sow when plants senesce. Sow in fall or early spring. Protect from predation until established.

Achnatherum hymenoides has showy flowers and seed heads that shimmer in the heat of dry inland areas. This attractive perennial bunchgrass grows to 2 ft. tall and produces open, branched clusters of wind-pollinated flowers that develop into a grain that can be eaten like rice. These seeds and the whole plant provide an important food source for wildlife in dry areas. It survives drought conditions by surrounding its roots with a rhizosheath, a microenvironment that harbors nitrogen-fixing organisms, which it creates with soil and mucilaginous secretions. This is a beautiful bunchgrass for gardens east of the Cascades. Some references classify this plant as *Eriocoma hymenoides*.

Indian ricegrass is a distinctive bunchgrass adapted to growing in dry, sandy conditions.

Carex geyeri • Cyperaceae

Elk sedge

HABITAT/RANGE Dry, open forests, woodlands, and grassy slopes. British Columbia to California and east to the Rocky Mountains at moderately low to high elevations. Grows mainly in and east of the Cascade Mountains.

SEASONAL INTEREST Flowers in spring to late summer relative to elevation. Evergreen.

WILDLIFE VALUE Provides cover and nesting materials. Butterfly host plant. Browsed by elk and other mammals, especially in early spring, less palatable to deer.

CULTIVATION Shade to part sun and seasonally moist, well-drained soil. Water to establish. Drought tolerant once established. Spreads slowly by rhizomes. Use as an evergreen groundcover in dry woodlands and shady xeric gardens. Mulch lightly.

While many sedges are found growing in wet places, elk sedge prefers drier upland habitat. This tough evergreen perennial is wind pollinated with flowering spikes that grow to around 1 ft. tall, though usually shorter, and long, leathery, grasslike leaves that lay in tufts on the ground. It creates a lovely drought-tolerant groundcover in eastside forests and woodlands. There are other sedges native to the region that also grow in dry upland habitats; use locally adapted species.

Elk sedge produces short, erect flowering spikes in spring, but the long, prostrate tufts of grassy evergreen leaves create waves of green in dry forest understories year-round.

Carex obnupta • Cyperaceae

Slough sedge

HABITAT/RANGE Wetlands, pond margins, riparian areas, wet meadows, and ditches. Northern British Columbia to California at low elevations. Grows mainly west of the Cascade Mountains.

SEASONAL INTEREST Flowers midspring to midsummer. Attractive seed heads. Evergreen in mild winters.

WILDLIFE VALUE Butterfly host plant. Seeds eaten by birds, waterfowl, and other wildlife. Provides forage, cover, and nesting materials. Helps provide healthy habitat for aquatic animals.

CULTIVATION Sun to part shade and wet soil. Can grow in shallow standing water; prefers freshwater but also grows in brackish coastal areas. Plant in a consistently wet site, otherwise provide supplemental water. Spreads vigorously by rhizomes in ideal conditions. Great for riparian areas, pond edges, bioswales, and wetland restoration.

Slough sedge's reliable structure and its many benefits to water quality and wildlife make it a great addition to riparian areas and wetlands west of the Cascades. This moisture-loving, rhizomatous perennial creates grasslike tufts that grow 2 to 4 ft. tall. The dark inflorescences provide some interest and are long lasting, first bearing small, wind-pollinated flowers and then seeds. Lyngbye's sedge (*Carex lyngbyei*) is similar and grows along the coast in salt marshes and wetlands. There are many species of sedge native to the Pacific Northwest.

Slough sedge is an attractive wetland plant that improves water quality.

Deschampsia cespitosa • Poaceae

Tufted hairgrass

HABITAT/RANGE Moist prairies, shorelines, tidal marshes, forests, subalpine meadows, alpine ridges, and disturbed areas. Alaska to California and east to the Atlantic Coast at low to high elevations. Widespread. Circumboreal. Grows on both sides of the Cascade Mountains.

SEASONAL INTEREST Flowers late spring to late summer depending on location.

WILDLIFE VALUE Butterfly host plant. Wind pollinated. Seeds eaten by birds. Provides cover for small animals, especially ground-nesting birds. Browsed by various species of wildlife.

CULTIVATION Sun to part shade and moist but well-drained, humus-rich soil. Grows in a variety of soil types. Water to establish and continue to provide supplemental water as needed. Tolerates seasonally dry conditions if properly sited. Great for butterfly gardens. Can be grown in containers. Mulch.

Tufted hairgrass is a popular native bunchgrass for both formal gardens and restoration plantings. It looks lovely either alone as a height accent in perennial beds or planted in swaths where the collective shimmer of its purplish to sandy colored inflorescences has a more pronounced effect. The basal foliage creates a tidy clump below the 2 to 4 ft. tall flowering stems. This is a widespread species; using locally sourced seed and plant material will provide gardeners with plants best suited to their local conditions. Slender hairgrass (*Deschampsia elongata*) is another attractive species that is widespread in the region.

Deschampsia cespitosa is a beautiful bunchgrass that produces tall stalks of shimmery flowers.

Elymus elymoides • Poaceae

Squirreltail

HABITAT/RANGE Widespread and variable. Grasslands, shrub-steppe, open woods, rocky slopes, coastal areas, and alpine areas. British Columbia to California and east to the Great Plains at low to high elevations. Grows on both sides of the Cascade Mountains.

SEASONAL INTEREST Flowers late spring to summer relative to elevation. Interesting seed heads.

WILDLIFE VALUE Host plant for the Nevada skipper butterfly. Provides cover and nesting materials for insects and wildlife. Browsed by a variety of species including deer, rabbits, and squirrels. More palatable in winter and spring before long awns develop.

CULTIVATION Sun to light shade and seasonally moist to dry, well-drained soil. Water to establish. Drought tolerant once established. This species consists of a few subspecies; use plants and seed from local sources when possible. A short-lived perennial; allow plants to reseed. Use in sunny xeric gardens, dry meadows, and rock gardens. Useful for restoration projects in arid areas.

This drought-tolerant perennial bunchgrass seems to have a bit more personality than most. It certainly has no problem showing off in hot, dry places with its flamboyant inflorescences boasting long awns and an iridescent purple hue. Squirreltail can reach 2 ft. tall, though it is usually shorter, and is a good choice for gardens east of the Cascades where it looks lovely tucked into a rock garden or planted in a dry, butterfly-friendly landscape.

Squirreltail is a showy and very drought-tolerant bunchgrass.

Festuca idahoensis • Poaceae

Idaho fescue

HABITAT/RANGE Grasslands, open forests, meadows, rocky slopes, and shrub-steppe. British Columbia to California and east to the Great Plains at low to high elevations. Grows mostly east of the Cascade Mountains.

SEASONAL INTEREST Flowers late spring to midsummer relative to elevation. Attractive seed heads.

WILDLIFE VALUE Butterfly host plant. Wind pollinated yet bees may collect pollen. Provides cover and nesting materials for birds, bees, and wildlife. Seeds eaten by birds. Valuable forage for many mammals. Usually ignored by deer.

CULTIVATION Sun to part shade and well-drained soil. Grows in a variety of soil types. Water to establish. Drought tolerant once established. Prefers some afternoon shade in hot, dry areas. Grows easily from seed. Sow in fall or early spring. Useful for erosion control. Plant in meadowscapes, sunny perennial beds, rock gardens, and habitat plantings. Does well in containers. Mulch lightly.

Idaho fescue is loved by gardeners for its glaucous hues and tidy, tufted habit. It brings long-lasting interest to the garden and provides a perfect contrast to shrubs and flowers with its green to bluish green leaves. Its flowering stems can grow up to 3 ft. tall and sway gracefully in the breeze over the dense basal foliage. Best of all, this drought-tolerant perennial provides food and shelter for birds and butterflies while it beautifies the landscape. There are many cultivars of this popular bunchgrass; choose plants wisely. Roemer's fescue (*Festuca roemeri*) is a good choice for gardens west of the Cascades.

Idaho fescue is an attractive drought-tolerant bunchgrass.

Juncus effusus subsp. *pacificus* • Juncaceae

Pacific rush

HABITAT/RANGE Wetlands, riparian areas, wet meadows, ditches, and coastal areas. Vancouver Island south through western Washington and Oregon to California, farther east in the Columbia River Gorge. Disjunct populations elsewhere. Low to mid elevations. Grows mainly west of the Cascade Mountains.

SEASONAL INTEREST Evergreen.

WILDLIFE VALUE Attracts beneficial insects. Seeds eaten by birds and small mammals. Provides cover and nesting habitat for waterfowl and wetland wildlife. Removes pollutants from water and soil. Deer resistant.

CULTIVATION Sun to part shade and moist to wet soil. Prefers consistent moisture. Tolerates flooding and shallow standing water. Grows in a variety of soil types. Spreads by seed and short rhizomes forming large clumps, which can be divided. Nitrogen fixing. Useful for riparian area stabilization and improving water quality. Plant around the edges of ponds or water features, in rain gardens, bioswales, and irrigated gardens. Can be grown in containers. Mulch.

Pacific rush provides a striking accent for wet sites and irrigated gardens. This rhizomatous evergreen perennial creates dense clumps of dark green cylindrical stems that grow 2 to 4 ft. tall. Clusters of wind-pollinated flowers bloom in summer near the tips of the stems. Beneficial to an array of wildlife, it creates habitat while improving water and soil conditions. *Juncus effusus* is a cosmopolitan plant with multiple subspecies. Exotic subspecies can be found growing wild in the region and are commonly sold by nurseries, sometimes mistakenly as natives. Choose plants wisely and plant our native subspecies, subsp. *pacificus*.

Pacific rush is a stout evergreen wetland plant.

Koeleria macrantha • Poaceae

Junegrass

HABITAT/RANGE Dry meadows, grasslands, prairies, open forests, oak savannas, shrub-steppe, and subalpine ridges. Widely distributed across North America at low to high elevations. Grows on both sides of the Cascade Mountains.

SEASONAL INTEREST Flowers late spring to summer. Attractive seed heads.

WILDLIFE VALUE Butterfly host plant. Wind pollinated yet bees may collect pollen. Seeds eaten by birds. Provides cover and nesting materials for birds, bees, and wildlife. Browsed by elk and deer.

CULTIVATION Sun to light shade and seasonally dry, well-drained soil. Water to establish. Drought tolerant once established. Grows easily from seed. Sow in fall or early spring. Cut back seed heads to deter spreading if needed. Tolerates mowing. Plant in sunny meadows, xeric perennial beds, rock gardens, butterfly gardens, and parking strips. Grows well in containers. Mulch lightly.

Junegrass is a popular, drought-tolerant, perennial bunch-grass that provides contrast and structure in the garden while benefiting a diversity of wildlife. Showy in bloom and fruit, the panicles of greenish to purplish flowers become dense and spikelike as seed develops. Growing to 2 ft. tall, sometimes taller, with a compact, clumping habit, the leaves are mostly basal and green to grayish green. Junegrass stalks and seed heads turn golden by late summer, providing color and contrast through the fall. This is a widely distributed species that varies due to adaptation to local climates. Use plant material propagated from local sources when possible.

Junegrass has beautiful flowers and seed heads that provide long-lasting interest in sunny gardens.

Leymus cinereus • Poaceae

Great Basin wildrye

HABITAT/RANGE Sunny, dry areas, grasslands, shrub-steppe, streamsides, and roadsides. Widespread across central and western North America at low to mid elevations. Grows east of the Cascade Mountains.

SEASONAL INTEREST Flowers in summer. Attractive seed heads.

WILDLIFE VALUE Butterfly host plant. Provides cover and nesting material for birds, insects, and mammals. Browsed by elk, deer, and small mammals.

CULTIVATION Sun to light shade and seasonally moist to dry, well-drained soil. Deeply rooted; does not like shallow soils. Grows in a variety of soil types. Water to establish. Drought tolerant once established. Creates dense clumps with growth points up to 1 ft. above the roots; do not cut back beyond that point. Slow to establish. Useful as a garden accent, for screening, and windbreaks. Mulch lightly.

Tall, robust grasses are popular in landscaping for the structural attraction and screening they provide. Sadly, exotic ornamentals such as pampas grass have escaped cultivation and become invasive. Great Basin wildrye offers a regionally appropriate alternative. Growing to a height of 6 ft. tall or more and spreading up to 3 ft. wide, this long-lived perennial bunchgrass is one of the largest in the region. Great Basin wildrye has a tightly clumping habit and grayish green leaves. Flowers and seed heads form dense spikes that can be used in flower arrangements. Best suited for sunny gardens east of the Cascades.

Leymus cinereus is a very large, drought-tolerant bunchgrass that makes a bold statement in gardens.

Schoenoplectus tabernaemontani • Cyperaceae

Softstem bulrush

HABITAT/RANGE Riparian areas, wetlands, marshes, and ditches. Widespread throughout North America, primarily at low elevations. Grows on both sides of the Cascade Mountains.

SEASONAL INTEREST Flowers late spring to late summer depending on location. Provides structural interest.

WILDLIFE VALUE Wind pollinated. Provides cover, nesting material, and nesting habitat for waterfowl, amphibians, reptiles, insects, and mammals. Seeds eaten by birds. Improves water quality.

CULTIVATION Full sun and shallow standing water or wet soil. Needs consistent moisture. Emergent in water up to 3 ft. deep. Spreads by seed and vigorous rhizomes. Plants can be grown in submerged pots to control spreading. Useful for soil stabilization in riparian areas. Plant along ponds, rivers, water features, and in bioswales.

Softstem bulrush, also called tule, is a robust wetland plant with round, pithy stems that quickly grows 3 to 9 ft. tall, making it useful as a screen along ponds and waterways. It spreads by thick rhizomes, creating dense colonies that hold riparian soils and buffer wave action, which helps create healthy riparian habitat. The flowers grow in loose clusters at the tips of the stems and produce seeds that are eaten by waterfowl and other birds. This water-loving perennial has an attractive form that is pleasing in the landscape. Previously known as *Scirpus validus*. Hardstem bulrush (*Schoenoplectus acutus*), also called tule, is taller and slightly more common.

Softstem bulrush is a tall, vigorous wetland plant that stabilizes riparian soils.

Typha latifolia • Typhaceae

Broadleaf cattail

HABITAT/RANGE Wetlands, riparian areas, coastal marshes, and ditches. Widespread throughout North America at low to mid elevations. Grows on both sides of the Cascade Mountains.

SEASONAL INTEREST Flowers in summer. Seed heads persist through winter. Fall color.

WILDLIFE VALUE Seeds eaten by waterfowl. Plants browsed by waterfowl and mammals. Hummingbirds collect seed fluff for nesting material in spring. Provides cover, nesting materials, and nesting habitat for birds, insects, and aquatic animals. Improves water quality.

CULTIVATION Sun to part shade and shallow standing water or saturated soil. Needs consistent moisture but tolerates some seasonal drought once established. Spreads aggressively by seed and rhizomes. Useful for soil stabilization, improving water quality, and creating wildlife habitat in riparian areas. Plant around ponds, in bioswales, and wet areas where it is free to spread.

Broadleaf cattail is an attractive wetland plant that provides multiple benefits to riparian ecosystems and wildlife. Growing 3 to 10 ft. tall, this emergent perennial has broad, grasslike leaves and dense spikes of flowers. The upper part of the flower spike consists of male flowers that eventually wither, leaving the darker female flowers to develop into a tightly packed seed head that persists through the winter and stands out against the golden fall foliage. Cattails have multifaceted uses for wildlife as well as people; they improve water quality and can be used as a cut flower, fiber, and food. Exotic species of cattail are found in the region. Choose plants wisely.

Broadleaf cattail is a common wetland plant that performs multiple functions in the ecosystem.

Ferns

Adiantum aleuticum • Pteridaceae

Western maidenhair fern

HABITAT/RANGE Shady, moist sites, forests, riparian areas, wet cliffs, and seeps. Alaska to California and east to Montana at low to high elevations, with disjunct populations elsewhere. Grows on both sides of the Cascade Mountains, though primarily west.

SEASONAL INTEREST Attractive foliage. Winter deciduous.

WILDLIFE VALUE Provides cover and nesting materials for birds and small animals. Deer resistant.

CULTIVATION Full to part shade and moist to wet, humus-rich soil. Grows in a variety of soil types, preferring ones rich in organic matter. Water to establish and continue to provide supplemental water as needed. Spreads slowly by rhizomes. Plant in wet rocky sites, shady riparian areas, rain gardens, and moist woodland gardens. Easy to grow if provided with sufficient shade and moisture. Can be grown in containers. Mulch.

Western maidenhair fern is a captivating perennial that grows 1 to 2 ft. tall from short rhizomes and has black stems that contrast its vibrantly green, palmately and horizontally arranged leaflets. Tucked among mossy rocks in moist, shady sites, it amplifies the level of lushness in the landscape with its lacy foliage. Plant it near a water source where it can provide a ferny umbrella of cover for birds and small animals darting in to take a drink. *Adiantum aleuticum* was previously considered a variation of *A. pedatum*, which is native to eastern North America. Exotic species of maidenhair fern, as well as hybrids and rare variations of native species are sold in nurseries. Choose plants wisely.

Western maidenhair fern has a gorgeous structure.

Athyrium filix-femina subsp. *cyclosorum* • Athyriaceae

Lady fern

HABITAT/RANGE Moist forests, riparian areas, bogs, and meadows. Alaska to California and east to the Great Plains at low to mid elevations. Grows on both sides of the Cascade Mountains.

SEASONAL INTEREST Attractive foliage. Winter deciduous.

WILDLIFE VALUE Provides cover and nesting materials. Sometimes browsed by elk and deer but considered deer resistant.

CULTIVATION Full to part shade and moist to wet, humus-rich soil. Water to establish and continue to provide supplemental water as needed. Spreads by spores and rhizomes in ideal conditions; not suitable for small spaces. Persists until frost but may look a bit ragged by end of summer. Spent fronds can be removed but leaving them protects plants in winter and provides shelter and nesting material for wildlife. Plant in moist woodland gardens and shady riparian areas. Mulch.

Athyrium filix-femina turns moist, shady woodlands into Jurassic Park with its lush foliage. A robust perennial, it can grow 5 ft. tall or more in ideal conditions, with mature plants spreading 3 to 7 ft. wide in a vaselike form. However large and in charge it seems, this fern has a delicate side expressed in its lacy foliage and fragile stems. It is a wide-ranging species with various subspecies. Many cultivars commonly available in nurseries have been derived from European subspecies; choose plants wisely and plant true natives. Spreading wood fern (*Dryopteris expansa*) is similar but only grows to about 3 ft. tall.

Lady fern is capable of filling in the forest understory.

Cystopteris fragilis • Cystopteridaceae

Fragile fern

HABITAT/RANGE Rocky sites, forest openings, and cliffs. Alaska to California and across northern North America at low to high elevations. Widespread, found worldwide. Grows on both sides of the Cascade Mountains.

SEASONAL INTEREST Attractive foliage. Summer deciduous.

WILDLIFE VALUE Provides nesting materials for small animals. Deer resistant.

CULTIVATION Part shade and moist to seasonally dry, well-drained, rocky soil. Water to establish. Drought tolerant once established if properly sited. Typically goes dormant in summer and reemerges in fall. May stay green through summer if adequate moisture is available. Spreads slowly by rhizomes. Plant in partly shaded rock gardens or tuck into rock walls. Does well in containers.

If you thought ferns were only for gardens with moist, dense shade, think again. Fragile fern is a sweet little species more adapted to growing in drier sites than many of the region's more robust and iconic ferns. While still shade-loving, it can take a bit of sun and grows in the light shade of oak trees, rocky forest openings, and even talus slopes above timberline. With a dainty appearance, this resilient perennial grows up to 1 ft. tall from slowly creeping rhizomes. The foliage provides a verdant dash of color in the spring but in the heat of summer it tends to go dormant. Give it a spot where it will not be overcome by other plants as fragile fern is a lover, not a fighter.

Cystopteris fragilis is a small, drought-tolerant fern.

Polypodium glycyrrhiza • Polypodiaceae

Licorice fern

HABITAT/RANGE On trees, rocks, logs, and moist slopes from Alaska to California, and southeastern British Columbia to Idaho. Low elevations. Found mainly in and west of the Cascade Mountains.

SEASONAL INTEREST Attractive foliage. Summer deciduous.

WILDLIFE VALUE Provides nesting materials. Deer resistant.

CULTIVATION Full to part shade. Place in mossy nooks of trees or logs, in woodland gardens among mossy rocks, tucked into rock walls, and along moist banks. Prefers big-leaf maple and other deciduous trees. Water to establish and continue to provide supplemental water as needed. Tolerates seasonally dry conditions once established if properly sited. Goes dormant by summer, reemerging with fall rains. Spreads slowly by rhizomes. Mulch lightly.

Licorice fern is an epiphytic perennial that can benefit the trees on which it grows. It is often found on mature big-leaf maples, which can produce canopy roots into pockets of organic matter created by licorice ferns and other epiphytic flora such as mosses. This greatly increases the amount of water and nutrients available to the trees. Licorice fern also grows on the ground or over mossy rocks, reaching 1 ft. tall or more from thick rhizomes known for their sweet licorice smell and flavor. The ethical harvest and propagation of this plant is important. If you want to use it, grow it. Consider rescuing plants from pruned or cut trees destined for the chipper. Other less common species of *Polypodium* occur in the region.

Licorice fern grows on trees as well as on the ground.

Polystichum munitum • Dryopteridaceae

Western sword fern

HABITAT/RANGE Moist forests from Alaska to California and east to northern Idaho and Montana. Low to mid elevations. Grows on both sides of the Cascade Mountains but is particularly abundant west of the mountains.

SEASONAL INTEREST Attractive foliage. Evergreen.

WILDLIFE VALUE Provides cover and nesting materials for many species. Important food source for elk, mountain beaver, and other wildlife. Browsed by deer, although generally considered deer resistant.

CULTIVATION Full to part shade and moist to seasonally moist, well-drained, humus-rich soil. Water to establish and continue to provide supplemental water as needed, especially in sunnier locations. Tolerates seasonally dry conditions in shady sites once established. Trim spent fronds in early spring if desired. Makes a nice accent plant in shaded planters. Mulch.

Western sword fern dominates the understory of moist conifer forests west of the Cascades where it provides ample shelter and forage for a wide array of wildlife. This robust evergreen perennial reaches 3 to 5 ft. tall and 3 to 4 ft. wide in ideal conditions, growing from a woody crown and rhizome. The leathery fronds are used in floral arrangements, although cutting green fronds can stunt the plant's growth, especially if more than half are taken. Other species of *Polystichum* found in the region, including some that are rare, look similar to western sword fern. Rock sword fern (*P. imbricans*) is more common in sunnier, drier sites.

Western sword fern is one of the region's most iconic plants, well-known for carpeting westside forests.

Struthiopteris spicant/Blechnum spicant • Blechnaceae

Deer fern

HABITAT/RANGE Moist forests, streambanks, coastal areas, and bogs. Alaska to California. Grows mainly west of the Cascade Mountains but also occurs from southeastern British Columbia into Idaho. Disjunct circumboreal distribution. Low to mid elevations.

SEASONAL INTEREST Attractive foliage and fiddleheads. Partly evergreen.

WILDLIFE VALUE Provides cover and nesting materials. Winter forage for elk, deer, rabbit, and other mammals, though generally considered deer resistant.

CULTIVATION Full to part shade and moist to wet, humus-rich soil. Likes acidic soils. Water to establish and continue to provide supplemental water as needed. Tolerates seasonally dry conditions once established if properly sited. Plant in shade gardens, coastal gardens, and moist woodlands. Can be grown in containers. Mulch.

The structure and color deer fern brings to moist, shady sites is sublime. From the unfurling of tight fiddleheads accented by reddish brown leafstalks to the light green or coppery tones of tender new fronds and the dark green shimmer of older fronds, this is a subtly but generously hued fern. This quality coupled with a tidy, clumping habit and distinct, pleasing form makes it an attractive addition to the landscape. Growing 1 to 3 ft. tall from short rhizomes, it is partly evergreen; the sterile fronds that extend gracefully from the plant are tough and evergreen, while the fertile fronds growing erectly in the center of the plant are deciduous.

Deer fern is a very attractive fern that is easy to grow if given adequate shade and moisture.

Shrubs

Amelanchier alnifolia • Rosaceae

Western serviceberry

HABITAT/RANGE Open forests, meadows, streambanks, and hillsides. Alaska to California and east to the Great Plains at low to high elevations. Grows on both sides of the Cascade Mountains.

SEASONAL INTEREST White flowers in spring to midsummer relative to elevation. Edible fruits in summer. Fall color. Deciduous.

WILDLIFE VALUE Attracts bees, butterflies, hummingbirds, and other pollinators. Butterfly and moth larval host. Provides cover and nesting sites. Fruits eaten by birds and other wildlife. Browsed by elk and deer.

CULTIVATION Sun to part shade and well-drained soil with some organic matter. Grows in a variety of soil types with adequate drainage. Water to establish. Drought tolerant once established if properly sited. Slow growing. Protect from deer until established. Broadleaf host for a rust it shares with juniper; avoid planting near juniper. Great for hedgerows, food forests, and backyard bird habitat. Mulch.

Western serviceberry, also called saskatoon, is beautiful in bloom, produces edible fruits, has colorful fall foliage, and supports pollinators and wildlife. Growing 5 to 20 ft. tall, sometimes taller, and 5 to 10 ft. wide depending on site conditions, this drought-tolerant deciduous shrub to small tree is slow to establish but eventually becomes a seasonal highlight with its profusions of white flowers and purplish blue fruits. The fruits look like blueberries but taste like apples and are a big attraction for birds. They can be used in pies, jams, and wine, but you will have to work that out with the birds. Many cultivars of this variable species are available; choose plants wisely and favor planting true natives.

Western serviceberry produces stunning white flowers and tasty fruits.

Arctostaphylos columbiana • Ericaceae

Hairy manzanita

HABITAT/RANGE Dry, sunny areas, open coniferous forests, forests edges, rocky slopes, thickets, and coastal areas from British Columbia to California. Low to mid elevations. Grows mainly in and west of the Cascade Mountains.

SEASONAL INTEREST White flowers late winter to spring. Rusty red fruits summer through winter. Evergreen.

WILDLIFE VALUE Important early-season pollen and nectar source for bees, butterflies, hummingbirds, and other pollinators. Buzz pollinated by bumble bees. Leaves used by leafcutter bees. Butterfly and moth host plant. Attracts beneficial insects. Fruits eaten by birds and other wildlife. Provides cover.

CULTIVATION Sun to light shade and well-drained, seasonally dry, slightly acidic soil. Drought tolerant. Water only to establish, providing occasional supplemental water for the first few years. Do not plant in irrigated gardens. Plant on a berm to shed excessive rainfall if needed. Start with small plants. Only prune dead wood. Has an ectomycorrhizal relationship that may benefit nearby conifers. Mulch.

Hairy manzanita has beautiful gray-green to bluish foliage and showy clusters of white to pink, urn-shaped flowers that attract hungry pollinators in early spring. Its rusty red, berrylike fruits feed wildlife and create long-lasting interest. A gorgeously contorted erect to spreading habit and peeling mahogany-red bark make this shrub a structural showpiece. One of the taller species of manzanita, it can grow 3 to 10 ft. tall and wide. This truly stunning evergreen shrub is a great choice for sunny, well-drained gardens from the coast to the mountains. The genus *Arctostaphylos* is particularly diverse in California.

Arctostaphylos columbiana is a beautiful evergreen shrub with attractive flowers, fruits, and foliage.

Arctostaphylos patula • Ericaceae

Green-leaf manzanita

HABITAT/RANGE Dry, open forests, mountain chaparral, and open slopes. Grows in mountainous areas from southern Washington through California and east to Colorado. Disjunct in north central Washington and Montana. Moderately low to mid elevations, sometimes higher.

SEASONAL INTEREST Pink flowers late winter to early summer relative to elevation. Brownish red fruits summer through winter. Evergreen.

WILDLIFE VALUE Early-season food source for bees, butterflies, moths, hummingbirds, and other pollinators. Buzz pollinated by bumble bees. Leaves used by leafcutter bees. Butterfly and moth host plant. Fruits eaten by birds and other wildlife. Provides cover.

CULTIVATION Sun to light shade and well-drained, seasonally dry, slightly acidic soil. Drought tolerant. Water only to establish, providing occasional supplemental water for the first few years. Do not plant in irrigated gardens. Plant on a berm to shed excessive rainfall if needed. Start with small plants. Only prune dead wood. Has mycorrhizal relationships. Nursery availability may be limited; difficult to propagate. Mulch.

Arctostaphylos patula begins to develop its inflorescences long before winter so it is ready to bloom at the first hint of spring's return, attracting butterflies and early-season pollinators with its hanging clusters of pink, urn-shaped flowers. Bringing both beauty and biodiversity to the landscape, this evergreen shrub grows 3 to 7 ft. tall and 3 to 9 ft. wide with thick, leathery leaves and peeling bark that reveals a smooth reddish underlayer on its contorted branches. Its brownish red, thick-skinned, berrylike fruits are attractive and feed wildlife. A mountain-dwelling species, it is adapted to cold and snowy winter climates.

Green-leaf manzanita brings structure and beauty to the garden throughout the year.

Arctostaphylos uva-ursi • Ericaceae

Kinnickinnick

HABITAT/RANGE Dry forests, rocky hillsides, coastal bluffs, sandy areas, and subalpine meadows. Widespread and circumboreal. Found throughout the Pacific Northwest, across Canada and the northern United States to the Atlantic Coast at low to high elevations.

SEASONAL INTEREST Pink flowers early spring to summer relative to elevation. Red fruits may persist through winter. Evergreen.

WILDLIFE VALUE Attracts bees, butterflies, moths, hummingbirds, and other pollinators. Buzz pollinated by bumble bees. Butterfly and moth host plant. Attracts beneficial insects. Fruits eaten by birds and other wildlife. Occasionally browsed by deer.

CULTIVATION Sun to part shade and well-drained, infertile, acidic soil. Does well in sandy or gritty soil. Water to establish. Drought tolerant. Prefers seasonally dry conditions once established, do not plant in regularly irrigated gardens. Has mycorrhizal relationships. Roots do not like disturbance. Useful for erosion control and an excellent groundcover for xeric gardens, rock gardens, pollinator gardens, and parking strips. Mulch lightly.

Kinnickinnick is much loved as a low-maintenance, drought-tolerant, evergreen groundcover. Hummingbirds and other pollinators are attracted to its pink, urn-shaped flowers that become showy, red, berrylike fruits. Growing to around 6 in. tall, the trailing woody stems create dense mats of leathery leaves that, while evergreen, can turn reddish purple in fall and winter. Many cultivars and exotic selections of this low-growing shrub are available in nurseries; choose plants wisely. Pinemat manzanita (*Arctostaphylos nevadensis*) is similar and grows in mountainous areas. *A. ×media* is a naturally occurring hybrid of *A. uva-ursi* and *A. columbiana*.

Kinnickinnick is a popular evergreen groundcover that benefits wildlife.

Artemisia tridentata • Asteraceae

Big sagebrush

HABITAT/RANGE Dry plains, shrub-steppe, foothills, and mountain slopes. Widespread and variable. Found throughout western North America at low to high elevations depending on the subspecies. Grows east of the Cascade Mountains.

SEASONAL INTEREST Yellow flowers late summer to fall. Silvery gray evergreen foliage.

WILDLIFE VALUE A keystone species vital to a diversity of wildlife including songbirds, birds of prey, sage grouse, insects, reptiles, and mammals. Provides food, cover, and nesting sites. Moth host plant. Attracts beneficial insects. Wind pollinated. Deer resistant.

CULTIVATION Full sun and dry, well-drained soil. Likes to grow in sandy or rocky soils. Water to establish. Drought tolerant once established. Has mycorrhizal relationships. Grows best east of the Cascades. Plant in hot, sunny sites and xeric gardens. Useful for erosion control, windbreaks, and hedgerows in arid areas.

Big sagebrush defines the landscape across large parts of the inland west and creates an oasis for biodiversity in seemingly harsh and inhospitable habitat. This rugged shrub grows anywhere from 3 to 6 ft. tall and wide, sometimes more. It has silvery gray, aromatic evergreen foliage and individually inconspicuous, but collectively showy, yellow flowers. The foliage and thick-trunked structure create a beautiful year-round focal point in xeric gardens; however, the volatile oils that make it so wonderfully fragrant also make it highly flammable and homeowners may want to distance it from buildings. *Artemisia tridentata* is a variable species that is comprised of numerous subspecies; use locally sourced or regionally appropriate plant material.

Big sagebrush is a well-known, drought-tolerant, evergreen shrub with intensely fragrant, silvery gray foliage.

Berberis aquifolium • Berberidaceae

Tall Oregon grape

HABITAT/RANGE Open woodlands and thickets. Southern British Columbia to California and east to Montana at low to mid elevations. Grows on both sides of the Cascade Mountains. Oregon state flower.

SEASONAL INTEREST Yellow flowers in spring. Edible blue berries in summer. Evergreen.

WILDLIFE VALUE Attracts bees, butterflies, hummingbirds, and other pollinators. Butterfly and moth host plant. Attracts beneficial insects. Berries eaten by birds and mammals. Provides cover and nesting sites. Occasionally browsed by deer and elk but considered deer resistant.

CULTIVATION Adaptable. Sun to part shade and well-drained, acidic soil. Prefers afternoon shade, especially in hot areas. Water to establish. Drought tolerant once established but will grow in irrigated sites. Spreads by rhizomes. Does not transplant well. Prune after flowering to shape if needed. Useful for hedges and screening. Plant in pollinator gardens and food forests. Mulch.

Tall Oregon grape is a popular landscaping plant with benefits for both gardeners and wildlife. It grows anywhere from 2 to 10 ft. tall and 3 to 6 ft. wide with spiny, hollylike leaves that emerge delicate and bronzy but become dark green, tough, and glossy. Although hardy, sun and cold temps can cause leaves of this evergreen shrub to turn shades of red and purple. Its showy clusters of small, yellow, daffodil-like flowers are richly perfumed and favored by pollinators, and its edible blue berries are loved by birds and can be used for jam. A great plant for hedgerows. Often classified as *Mahonia aquifolium*.

Tall Oregon grape is a drought-tolerant evergreen shrub with fragrant flowers that attract pollinators and fruits that feed birds.

Berberis nervosa • Berberidaceae

Dull Oregon grape

HABITAT/RANGE Forests and forest edges. Southern British Columbia to California and scattered locations east to northwestern Montana. Low to mid elevations. Grows mainly in and west of the Cascade Mountains.

SEASONAL INTEREST Yellow flowers in spring to early summer. Blue berries in summer. Evergreen.

WILDLIFE VALUE Attracts bees, butterflies, hummingbirds, and other pollinators. Butterfly and moth larval host. Attracts beneficial insects. Berries eaten by birds and mammals. Provides cover. Occasionally browsed by deer and elk but considered deer resistant.

CULTIVATION Full shade to part sun and moist to seasonally moist, well-drained, humus-rich soil. Prefers slightly acidic soils. Water to establish. Drought tolerant once established if properly sited but can grow in irrigated sites. Slow to establish. Rhizomatous. Does not transplant well. Plant in shady sites, woodland gardens, food forests, and under conifers. Mulch.

Dull Oregon grape is a handsome, shade-loving evergreen groundcover that grows up to 2 ft. tall. The dark green leathery leaves have spiky edges and are pinnately compound, radiating out from erect stems. Older leaves turn shades of red and purple, and young leaves have reddish hues making the foliage of this evergreen shrub quite colorful at various times of the year. The fragrant yellow flowers attract pollinators before becoming grapelike clusters of blue berries that can be used in jams or left to feed the birds. Often classified as *Mahonia nervosa*.

Dull Oregon grape brings year-round structural appeal to shady sites.

Berberis repens • Berberidaceae

Creeping Oregon grape

HABITAT/RANGE Open forests, shrublands, and grasslands. Southern British Columbia to California and east beyond the Rocky Mountains at low to high elevations. Grows east of the crest of the Cascade Mountains.

SEASONAL INTEREST Yellow flowers in spring to early summer. Blue berries in summer. Evergreen.

WILDLIFE VALUE Attracts bees, butterflies, and other pollinators. Butterfly and moth host plant. Attracts beneficial insects. Berries eaten by birds and mammals. Provides cover for small animals. Occasionally browsed by deer and elk but considered deer resistant.

CULTIVATION Sun to shade and well-drained soil. Prefers part shade and slightly acidic soils with some organic matter. Water to establish. Drought tolerant once established but can handle occasional watering. Slow growing. Spreads by rhizomes and stolons. Use as a groundcover in dry woodlands, pollinator gardens, rock gardens, and parking strips. Mulch lightly.

Berberis repens is a drought-tolerant, low-growing species of Oregon grape that can handle both sun and shade. The evergreen leaves are pinnately compound and leathery with spiny teeth, often becoming tinged red and purple in cold weather or excessive sun. The fragrant yellow flowers attract bees and butterflies and produce clusters of edible blue berries. Not your typical shrub, this creeping groundcover keeps a low profile and only grows around 1 ft. tall, sometimes taller. All these qualities make it popular with gardeners and landscapers. An especially good choice for gardens east of the Cascades. Hybridizes with *B. aquifolium* and is often classified as *Mahonia repens* or a variation of *B. aquifolium*.

Creeping Oregon grape is a low-growing shrub with bright yellow flowers.

Ceanothus integerrimus • Rhamnaceae

Deerbrush

HABITAT/RANGE   Dry woodlands and forests, open areas, chaparral, rocky slopes, and disturbed areas. Southern Washington to Baja California at low to mid elevations. Grows on both sides of the Cascade Mountains.

SEASONAL INTEREST   Purplish blue, white, or pink flowers in late spring to midsummer, sometimes sporadically into early fall. Deciduous.

WILDLIFE VALUE   Attracts bees, butterflies, and other pollinators. Important butterfly and moth host plant. Attracts beneficial insects. Seeds eaten by birds. Provides food, cover, and nesting sites. Browsed by deer.

CULTIVATION   Sun to part shade and seasonally dry, well-drained soil. Water only to establish. Drought tolerant once established. Sometimes finnicky and difficult to establish. Prune after flowering to shape if needed. Good for hedgerows and xeric butterfly gardens. Protect from deer until established. Mulch.

Deerbrush, also called wild lilac, is a deciduous shrub that produces elongated clusters of fragrant purplish blue, white, or pink flowers. It grows 3 to 12 ft. tall and as much as 7 ft. wide. Adapted to fire, plants senesce after about 30 years if not "pruned" by low-intensity fires, which can cause plants to regenerate and seeds to germinate. Redstem ceanothus (*Ceanothus sanguineus*) ranges farther north, has white flowers, and grows similarly to deerbrush. Mountain balm (*C. velutinus*) is an important butterfly host plant and fragrant evergreen with creamy white flowers. Common buckbrush (*C. cuneatus*) is easy to grow and ranges from the Willamette Valley south. These aromatic, nitrogen-fixing, drought-tolerant shrubs all support pollinators and wildlife.

Deerbrush is a prolific bloomer that creates stunning displays of color and feeds a variety of pollinators.

Ceanothus thyrsiflorus • Rhamnaceae

Blue blossom

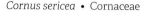

HABITAT/RANGE Chaparral, open slopes, and forests along the coast from central Oregon south through California. Low elevations. Grows west of the Cascade Mountains.

SEASONAL INTEREST Light to dark or purplish blue flowers in mid-spring to early summer, sometimes again in fall. Evergreen.

WILDLIFE VALUE Attracts bees, butterflies, and other pollinators. Butterfly and moth larval host. Attracts beneficial insects. Provides evergreen cover and nesting sites. Seeds eaten by birds and small mammals. Browsed by elk and deer.

CULTIVATION Sun to part shade and well-drained soil. Water to establish. Drought tolerant once established and easy to grow if properly sited. No summer water needed once established. Not very cold tolerant; may perish in harsh winters. Hardy to USDA zone 7b. An excellent evergreen for pollinator-friendly hedgerows, screening, and coastal gardens. Can be pruned after flowering to shape. Mulch.

Blue blossom turns heads with its copious flowers that range from light to dark or purplish blue. This evergreen shrub's only drawback is that it is somewhat tender and prefers mild, maritime climates. If you live where this showy shrub can survive the winter, you will find it to be fast growing and abundantly beautiful. It is variable in height, shape, and flower color. Taking the form of a low, mounding shrub or small tree, it grows 3 to 18 ft. tall or more and up to 9 ft. wide. There are many selections and hybrids available, providing gardeners with size and color options. The genus *Ceanothus* is particularly diverse in California.

Blue blossom is an evergreen shrub with treelike tendencies that blooms profusely.

Cornus sericea • Cornaceae

Red-osier dogwood

HABITAT/RANGE Riparian areas, wetlands, and moist forests. Widespread. Alaska to Mexico and across northern North America at low to mid elevations. Grows on both sides of the Cascade Mountains.

SEASONAL INTEREST White flowers in spring to summer, sometimes sporadically into fall. White to pale blue fruits. Fall color. Red twigs provide winter interest. Deciduous.

WILDLIFE VALUE Attracts bees, butterflies, and other pollinators. Butterfly and moth host plant. Attracts beneficial insects. Fruits eaten by birds, mammals, and fish. Provides cover and nesting sites. Plants browsed by deer, elk, beaver, and other wildlife.

CULTIVATION Sun to shade and moist to wet, humus-rich soil. Water to establish and continue to provide supplemental water as needed. Easy to grow and quick to establish if given adequate moisture. Tolerates flooding and some seasonal drought. Branch color more vibrant in sunny sites. Propagates easily by cuttings. Excellent for riparian areas, rain gardens, bioswales, wetlands, hedgerows, and backyard bird habitat. Useful for erosion control. Mulch.

Red-osier dogwood is a workhorse of a plant. It holds soils, shades waterways, provides habitat, and feeds an array of wildlife from butterflies to birds, and even fish. It grows quickly in moist soils and provides year-round interest with its flat-topped clusters of creamy flowers, berrylike fruits, fall color, and reddish twigs that are surprisingly showy in winter. Depending on site conditions this deciduous shrub grows anywhere from 6 to 18 ft. tall and 6 to 10 ft. wide. Sources vary in its classification; *Cornus stolonifera* and *C. occidentalis* are synonyms for *C. sericea*.

Red-osier dogwood is a beautiful, moisture-loving shrub that benefits wildlife.

Corylus cornuta • Betulaceae

Beaked hazelnut

HABITAT/RANGE Forest openings and edges, thickets, and damp, rocky slopes. British Columbia to California, mostly west of the Cascade Mountains, with one subspecies occurring farther east and to the Atlantic Coast. Low to mid elevations.

SEASONAL INTEREST Catkins midwinter to early spring. Edible nuts. Fall color. Deciduous.

WILDLIFE VALUE Wind pollinated. Nuts eaten by wildlife. Host for the gorgeous Polyphemus moth. Attracts beneficial insects. Provides cover and nesting sites. Browsed by wildlife, infrequently by deer.

CULTIVATION Sun to shade and moist to seasonally dry, well-drained, humus-rich soil. Tolerates clay if not saturated. Water to establish. Drought tolerant once established if properly sited. Suckers. May slowly form thickets in ideal conditions. A good plant for hedgerows, food forests, and wildlife habitat. Mulch.

Hazelnuts, or filberts, are known for their edible nuts, which are surrounded by a rough sheath and favored by squirrels who are quick to harvest them. Our native species may not produce as copiously as cultivated European filberts, yet their nuts are still highly desirable. This deciduous shrub to small tree grows anywhere from 5 to 25 ft. tall and 5 to 20 ft. wide depending on site conditions. It has small red female flowers and showy male catkins that bloom in late winter on naked branches before the leaves emerge. The wide, hairy leaves are a lovely shade of green and turn bright yellow in fall. Our native species is resistant to the feared eastern filbert blight; however, the import of *Corylus* spp. is restricted.

Beaked hazelnut is best known for its edible nuts.

Dasiphora fruticosa • Rosaceae

Shrubby cinquefoil

HABITAT/RANGE Rocky slopes, shrub-steppe, meadows, bogs, shorelines, and subalpine areas. Circumboreal. Alaska to California and east to northeastern North America at low to high elevations. Grows on both sides of the Cascade Mountains.

SEASONAL INTEREST Long blooming. Yellow flowers from late spring to early fall. Deciduous.

WILDLIFE VALUE Attracts bees, butterflies, and other pollinators. Many solitary bees specialize on woody species in the rose family. Butterfly host plant. Attracts beneficial insects. Seeds eaten by birds and small mammals. Provides cover and nesting materials. Browsed by deer.

CULTIVATION Sun to light shade and moist to seasonally dry soil. Prefers well-drained soils but grows in a variety of soil types including clay. Water to establish. Drought tolerant once established if properly sited. Plant in sunny spots, rock gardens, parking strips, hedgerows, and pollinator gardens. Mulch lightly.

Shrubby cinquefoil occurs in a variety of habitats and while it is drought tolerant once established, it also thrives with evenly moist soils making it a versatile plant in the landscape. It grows to around 3 ft. tall and wide with a dense, upright to prostrate habit and small, hairy, green to silvery gray leaves. This cold-hardy deciduous shrub can bloom throughout the summer, producing yellow, saucer-shaped flowers that attract butterflies and bees. A popular landscaping plant, numerous cultivars have been derived from it; choose plants wisely. Some still refer to this plant as *Potentilla fruticosa*.

Shrubby cinquefoil is a long-blooming shrub with showy yellow flowers.

Ericameria nauseosa • Asteraceae

Gray rabbitbrush

HABITAT/RANGE Open, dry areas, shrub-steppe. Widely distributed throughout the western United States and adjacent southern Canada at low to fairly high elevations. Grows mainly east of the Cascade Mountains.

SEASONAL INTEREST Yellow flowers in late summer to fall. Silvery gray foliage. Evergreen in mild winter climates.

WILDLIFE VALUE Important for butterflies, bees, and other pollinators. Fuels the southward migration of monarch butterflies. Attracts beneficial insects. Butterfly and moth host plant. Seeds eaten by birds. Provides cover and nesting sites. Not browsed heavily.

CULTIVATION Full sun and dry, well-drained soil. Water only to establish. Drought tolerant once established. Tolerates heat and poor soils. Cutting new growth back by half after flowering can help maintain a denser form. Great for xeric gardens, parking strips, and butterfly gardens east of the Cascades. Mulch lightly if at all.

Gray rabbitbrush is one of the few plants in the dry Intermountain West to bloom late in the growing season, which it does in abundance. Its blooms create a highway of yellow-flowered fueling stations for western populations of monarch butterflies migrating south. It is important to a wide array of pollinators and wildlife that depend on the food and shelter it provides. Usually remaining shorter than its maximum height and width of 6 ft., the silvery gray stems and leaves of this plant bring contrast to the garden and are striking against the yellow flowers. Green rabbitbrush (*Chrysothamnus viscidiflorus*) is similar and grows in the same habitat. Coyote brush (*Baccharis pilularis*) is a fall-blooming evergreen shrub that supports native pollinators and grows west of the Cascades.

Gray rabbitbrush is colorful in bud long before its bright yellow flowers bloom in fall.

Garrya elliptica • Garryaceae

Wavyleaf silk-tassel

HABITAT/RANGE Coastal areas from west central Oregon to Southern California. Mainly at low elevations, sometimes higher.

SEASONAL INTEREST Showy catkinlike flowers in winter. Purplish fruits produced by female plants. Evergreen.

WILDLIFE VALUE Wind pollinated. Fruits eaten by birds. Provides evergreen cover.

CULTIVATION Sun to part shade and well-drained soil. Water to establish. Drought tolerant once established, especially in coastal areas; supplemental water may be needed elsewhere. This plant is hardy to USDA zone 7b and is too tender for gardens in colder climates. Placing it in a protected spot out of prevailing winter winds may help it survive cold weather. Leaves can be burnt by both hot and cold temperatures. Good evergreen choice for a hedge or screen. Prune after flowering if needed. Mulch.

Wavyleaf silk-tassel is a popular landscape plant favored for its evergreen foliage and pendulous, catkinlike flowers in winter. Growing 7 to 16 ft. tall and wide, this shrub to small tree is dioecious, producing male and female flowers on separate plants. The blooms are long lasting and while the male flowers are a little showier, the female plants bear the grapelike clusters of hairy, purplish fruits that birds, but not humans, eat. The leaves are thick and leathery with wavy margins. The natural range of this evergreen shrub is limited to coastal areas with mild summer and winter climates; it is not appropriate for gardens in areas with more extreme conditions.

Wavyleaf silk-tassel is a showy winter bloomer that produces long, catkinlike flowers.

Garrya fremontii • Garryaceae

Fremont's silk-tassel

HABITAT/RANGE Mixed conifer forests, woodlands, chaparral, and rocky slopes from near the Columbia River Gorge in Washington to Southern California. Low to mid elevations. Grows in and on both sides of the Cascade Mountains.

SEASONAL INTEREST Catkinlike flowers in winter to spring. Blue fruits produced on female plants. Evergreen.

WILDLIFE VALUE Wind pollinated. Fruits eaten by birds and small mammals. Provides evergreen cover.

CULTIVATION Sun to part shade and well-drained soil. Water to establish. Drought tolerant once established if properly sited. Hardy to USDA zone 6a; more heat and cold tolerant than *Garrya elliptica*. A good choice for a hedge or screen. Prune after flowering if needed. Mulch.

Garrya fremontii is a handsome evergreen shrub that grows in the cold climates of the Cascades and blooms profusely in late winter to early spring. Dioecious, it produces pendulous, catkinlike male and female flowers on separate plants. Male flowers are slightly showier but, if there is a male nearby to pollinate it, female plants produce clusters of blue fruits that are edible to wildlife but not humans. The evergreen leaves are thick, leathery, and shiny on top. This drought-tolerant shrub grows to about 10 ft. tall and wide and looks beautiful year-round as a focal point in the landscape.

Fremont's silk-tassel is a showy evergreen shrub that blooms in winter to spring.

Salal is an attractive evergreen shrub with edible fruits.

Gaultheria shallon • Ericaceae

Salal

HABITAT/RANGE Forests and coastal areas from Alaska to Southern California. Low to mid elevations. Grows from the eastern base of the Cascade Mountains to the coast.

SEASONAL INTEREST White to pinkish flowers in spring to midsummer. Edible blue fruits. Evergreen.

WILDLIFE VALUE Attracts bees, butterflies, hummingbirds, and other pollinators. Butterfly and moth host plant. Attracts beneficial insects. Fruits eaten by birds and other wildlife. Provides evergreen cover and nesting sites. Winter browse for deer and elk.

CULTIVATION Shade and moist to seasonally moist, humus-rich soil. Tolerates sun but may suffer leaf burn. Prefers acidic soils rich in organic matter but tolerates poor soils. Water to establish. Drought tolerant once established if properly sited. Supplemental water may be needed in drier inland environments. Slow to establish but eventually grows vigorously. Rhizomatous. Typically does not transplant well. A choice plant for shady hedges, food forests, and wildlife habitat. Loves ample applications of mulch.

Salal is a staple of the coastal understory and with qualities like attractive evergreen foliage, interesting flowers that feed pollinators, and edible fruits, it should be a staple in westside gardens too. After an awkward establishment period this plant can grow robustly and continue to spread, reaching anywhere from 2 to 10 ft. tall depending on site conditions. The urn-shaped flowers are white to pink and turn into blue, berrylike fruits. The stems zigzag slightly between leathery leaves and the foliage of this evergreen shrub has long been favored by the floral industry for the sturdy greenery it provides.

Holodiscus discolor • Rosaceae

Oceanspray

HABITAT/RANGE Moist to dry forests and woodlands, coastal bluffs, rocky slopes, riparian areas, and disturbed areas. Widely distributed. British Columbia to Southern California and east to the Rocky Mountains. Low to mid elevations. Grows on both sides of the Cascade Mountains.

SEASONAL INTEREST White flowers in early to late summer depending on location. Some fall color. Deciduous.

WILDLIFE VALUE Attracts bees, butterflies, and other pollinators. Butterfly and moth larval host. Attracts beneficial insects. Birds eat seeds and glean insects from plants. Provides cover and nesting sites. Typically ignored but may be browsed by deer and elk.

CULTIVATION Sun to part shade and moist to seasonally dry soil. Tolerates a variety of soil types. Water to establish. Drought tolerant once established. Supplemental summer water in arid areas can keep plants looking lush. Fast growing. Spreads by seed. Prune after flowering to shape if desired but letting seed heads persist into winter provides food for birds. Good for xeric gardens, hedgerows, and butterfly gardens. Mulch.

Oceanspray is a flamboyant shrub that produces profusions of creamy flowers in cascading clusters from its branch tips. Don't let the name fool you, this deciduous shrub is quite drought tolerant. It grows quickly, reaching 3 to 15 ft. tall and wide depending on site conditions, with an erect to arching habit. A larval host for some of the most beautiful species of butterflies and moths in the region, oceanspray truly brings beauty and biodiversity to the landscape.

When in bloom it is obvious where oceanspray gets its common name, but this drought-tolerant shrub can thrive in areas far from the sea.

Lonicera ciliosa • Caprifoliaceae

Orange honeysuckle

HABITAT/RANGE Forested areas and thickets. Southern British Columbia to Northern California and east to northern Idaho and Montana at low to mid elevations. Grows on both sides of the Cascade Mountains.

SEASONAL INTEREST Orange flowers in late spring to midsummer. Red berries. Deciduous.

WILDLIFE VALUE Flowers favored by hummingbirds. Attracts butterflies, bees, and other pollinators. Attracts beneficial insects. Host plant for checkerspot butterflies. Fruits eaten by birds.

CULTIVATION Part shade and moist to seasonally moist soil. Water to establish. Drought tolerant once established if properly sited. Needs a trellis, nearby shrub, or tree to climb up, otherwise it creeps along the ground. Quick growing but not aggressive. Prone to powdery mildew, which does not harm the plant. A good plant for hummingbird gardens. Mulch.

Orange honeysuckle is a deciduous vine with large, orange, tubular flowers that are a magnet for hummingbirds. It climbs quickly up nearby shrubs and trees and pairs well with shrubs like mock orange (*Philadelphus lewisii*). The last set of opposite leaves at the stem tips are fused behind the clusters of flowers, and its fruits are translucent red berries, which disappear quickly as birds eat them. While fast growing, this woody vine has a sparse and slender habit and does not usually create a thick screen or have an aggressive presence in the garden. Well worth growing for its dazzlingly beautiful flowers and benefits for birds.

Orange honeysuckle is a woody vine with showy tubular flowers that beckon to hummingbirds.

Lonicera involucrata • Caprifoliaceae

Black twinberry

HABITAT/RANGE Moist forests, thickets, riparian areas, marshes, and subalpine areas. Widespread. Alaska to Mexico and east to the Rocky Mountains and beyond at low to high elevations. Grows on both sides of the Cascade Mountains.

SEASONAL INTEREST Yellow flowers in spring to late summer relative to elevation. Black berries with showy red bracts. Deciduous.

WILDLIFE VALUE A magnet for bumble bees and hummingbirds. Attracts butterflies and other pollinators. Attracts beneficial insects. Butterfly and moth host plant. Fruits eaten by birds and mammals. Provides cover and nesting sites. Typically ignored but may be browsed by deer and elk.

CULTIVATION Sun to shade and moist, humus-rich soil. Water to establish and continue to provide supplemental water as needed. Benefits from deep, infrequent summer irrigation. Fast growing if given sufficient moisture. Prefers afternoon shade in hot areas. A good choice for forest edges, moist hedgerows, rain gardens, riparian areas, and pollinator gardens. Easily pruned. Mulch.

Black twinberry is fast growing, reaching about 5 to 10 ft. tall and wide, sometimes more, with an erect to spreading habit. This leafy deciduous shrub produces pairs of yellow tubular flowers that are loved by bumble bees and hummingbirds. After pollination pairs of black berries surrounded by thick red bracts develop that are not edible to humans but relished by birds, making this a great addition to backyard habitat plantings. Two variations of *Lonicera involucrata* occur in the region; var. *ledebourii* is more common along the coast in Oregon and California and has leathery leaves and red-tinged flowers.

Black twinberry has yellow tubular flowers and interesting fruits.

Morella californica/Myrica californica • Myricaceae

Pacific wax myrtle

HABITAT/RANGE Coastal areas. Central Washington to Southern California at low elevations. Disjunct on the west coast of Vancouver Island. At-risk species of special concern in British Columbia.

SEASONAL INTEREST Inconspicuous flowers in spring. Warty, purplish fruits late summer or fall. Evergreen.

WILDLIFE VALUE Wind pollinated. Fruits eaten by birds. Provides evergreen cover. Fairly deer resistant.

CULTIVATION Sun to part shade and moist to seasonally dry, well-drained soil. Grows in a variety of soil types including sandy soils. Water to establish. Moderately drought tolerant once established, especially in coastal gardens. May need some shade and occasional supplemental water inland. Too tender for many areas; hardy to USDA zone 7a. An excellent evergreen for formal landscapes, screens, hedges, windbreaks, and bird habitat. Prune as needed. Mulch.

Pacific wax myrtle is a choice ornamental. This coastal evergreen shrub to small tree is beautiful year-round and has long, glossy leaves that are fragrant, especially when crushed. The flowers are inconspicuous, but the small, warty, purplish fruits feed birds and can be used to make natural dyes. It grows quickly, reaching 5 to 30 ft. tall and 5 to 20 ft. wide, and eventually becomes a small tree; however, it is easily pruned and maintained as an evergreen hedge. Pacific wax myrtle can grow in a variety of soil types and is able to tolerate infertile soils as it hosts nitrogen-fixing bacteria that help supply the plant with nutrients.

Pacific wax myrtle is a fragrant evergreen excellent for hedges, screens, and bird habitat.

Oemleria cerasiformis • Rosaceae

Osoberry

HABITAT/RANGE Woodlands, lowland forests, thickets, and riparian areas. Southern British Columbia to Southern California at low elevations, sometimes higher. Grows mainly west of the Cascade Mountains.

SEASONAL INTEREST White flowers in late winter to spring. Fruits ripen in summer. Deciduous.

WILDLIFE VALUE Early-season nectar source for pollinators like mason bees. Attracts bees, butterflies, hummingbirds, and other pollinators. Moth host plant. Fruits eaten by birds and mammals. Provides cover and nesting sites.

CULTIVATION Part shade to sun and moist to seasonally dry, humus-rich, well-drained soil. Water to establish and continue to provide supplemental water as needed. Drought tolerant once established if properly sited. Both a male and female needed to produce fruits. Fruit production higher with sun and adequate moisture. Suckering shrub; not appropriate for small spaces. Great for shady, lowland forest gardens, restoration plantings, and bird habitat. Mulch.

Before the broad-leaved deciduous trees of lowland forests have unfurled their greenery, osoberry produces hanging clusters of white flowers and tufts of cucumber-scented leaves that catch the sunlight and signal the coming of spring. These early-season flowers are a valuable food source for pollinators. Female plants develop small, plumlike stone fruits that ripen in summer. The colorful fruits go from beautiful shades of peach to a deep purplish blue and are loved by many birds. They are also edible to humans but bitter until fully ripe. Osoberry is a leafy deciduous shrub with an open, erect habit reaching 5 to 15 ft. tall and wide.

Osoberry is an early bloomer that produces colorful plumlike fruits that turn dark purplish blue when ripe.

Paxistima myrsinites • Celastraceae

Oregon boxwood

HABITAT/RANGE Forests, shaded sites, and open, rocky areas. British Columbia to California and east to the Rocky Mountains and northern Mexico. Low to high elevations. Grows on both sides of the Cascade Mountains.

SEASONAL INTEREST Small maroon flowers in spring to summer relative to elevation. Evergreen.

WILDLIFE VALUE Attracts bees and other pollinating insects. Provides cover. Browsed by deer, elk, grouse, and other wildlife.

CULTIVATION Part shade and moist to seasonally dry, well-drained soil. Tolerates sun. Fairly adaptable to different soil types; prefers slightly acidic, humus-rich soils that do not remain saturated. Water to establish. Drought tolerant once established. Can grow in dry shade under conifers. Slow to moderate growth rate depending on site conditions. A great choice for low hedges, borders, rock gardens, and forest gardens. Can be pruned to shape. Mulch.

Also called mountain lover, Oregon boxwood is a low-growing evergreen shrub commonly found in the understory of mountain forests. It is shade loving but can grow in the sun where it tends to be more compact. It is popular as a hedge plant and offers gardeners and landscapers a native alternative to commonly used nonnative species like European boxwood and Japanese holly. This erect to spreading shrub has a short stature, reaching 3 ft. tall and wide. The inconspicuous flowers bloom singly or in clusters along the branches at leaf axils. The serrate, evergreen leaves are small and leathery with new growth emerging in light green shades that contrast and complement the older, darker foliage.

Oregon boxwood is hardy and great for hedges.

Philadelphus lewisii • Hydrangeaceae

Mock Orange

HABITAT/RANGE Open woodlands, forest edges, riparian areas, rocky hillsides, and seasonally moist areas. Widespread from southern British Columbia to California and east to Montana at low to mid elevations. Grows on both sides of the Cascade Mountains. Idaho state flower.

SEASONAL INTEREST Fragrant white flowers late spring to midsummer relative to elevation. Brief fall color. Deciduous.

WILDLIFE VALUE Attracts bees, butterflies, and other pollinators. Favored by swallowtail butterflies. Attracts beneficial insects. Moth host plant. Seeds eaten by wildlife. Provides cover and nesting sites. Browsed by deer and elk.

CULTIVATION Adaptable; grows in a wide range of conditions. Sun to part shade and moist to seasonally dry soil. Water to establish. Drought tolerant once established if properly sited. Prefers humus-rich soils and occasional summer water. Grows taller in moist soils and blooms more profusely in sunny sites. Great for hedgerows and screening. Prune after flowering to shape if desired. Protect from deer until established. Mulch.

If there is a "must-have" list for native plants, mock orange is on it. This adaptable, drought-tolerant deciduous shrub establishes quickly and blooms profusely, covering itself in showy white flowers that fill the air with a sweet scent similar to orange blossoms. It grows as much as 10 ft. tall and almost as wide with an erect to arching form and dense branching habit. A great plant for pollinators and a popular landscaping plant with many cultivars available; avoid using double-flowered varieties as they can be confusing and inaccessible to pollinators.

Mock orange blooms profusely and is one of the most sweetly scented shrubs in the region.

Physocarpus capitatus • Rosaceae

Pacific ninebark

HABITAT/RANGE Moist forests, riparian areas, wetlands, and coastal marshes. Alaska to California and east to Idaho at low to mid elevations. More common west of the Cascade Mountains.

SEASONAL INTEREST White flowers late spring to early summer. Fall color. Deciduous.

WILDLIFE VALUE Attracts bees, butterflies, and other pollinators. Attracts beneficial insects. Butterfly and moth larval host. Seeds eaten by birds. Provides ample cover and nesting sites. Shades waterways and stabilizes riparian soils creating healthy habitat for aquatic and amphibious wildlife. Browsed by deer.

CULTIVATION Sun to part shade and moist, humus-rich soil. Water to establish and continue to provide supplemental water as needed. Creates a dense hedge or screen that can be pruned as needed. Plant in moist sites, hedgerows, irrigated gardens, rain gardens, and riparian areas. Mulch.

Pacific ninebark's rounded clusters of white flowers brighten wetlands, waterways, and moist sites in late spring. This deciduous shrub combines a pleasant arching habit with showy flowers, attractive foliage, and peeling bark, making it a desirable landscape plant. It grows quickly to anywhere from 5 to 13 ft. tall and 4 to 7 ft. wide. Mallow ninebark (*Physocarpus malvaceus*) looks similar, is more drought tolerant, and grows east of the Cascades. There are many cultivars of ninebark available, particularly of species native to eastern North America. Choose plants wisely, plant true natives, and avoid cultivars with dark red to purple foliage, which may be toxic to native insects.

Pacific ninebark is a beautiful flowering shrub with an arching habit that grows in moist areas.

Purshia tridentata • Rosaceae

Bitterbrush

HABITAT/RANGE Open, dry areas, shrub-steppe, and pine forests. Widely distributed from southern British Columbia south through California and east to just beyond the Rocky Mountains. Low to high elevations. Grows east of the Cascade Mountains.

SEASONAL INTEREST Pale yellow flowers in spring to midsummer relative to elevation. Deciduous.

WILDLIFE VALUE Important for pollinators and wildlife. Attracts bees, butterflies, and other pollinators. Attracts beneficial insects. Butterfly and moth host plant. Seeds eaten by birds and small mammals. Provides valuable cover for wildlife and birds including sensitive species like sage grouse. Vital forage for wildlife; browsed by deer and antelope.

CULTIVATION Full sun and well-drained, dry soils. Water only to establish. Drought tolerant once established. Plant in xeric gardens. Best suited for gardens east of the Cascades. Useful for erosion control, land reclamation, and rangeland restoration. Mulch lightly if at all.

Bitterbrush is an important part of the shrub-steppe plant community and a critical plant for wildlife in arid areas of western North America. This drought-tolerant deciduous shrub is nitrogen fixing and has small, trilobed leaves that have a grayish green cast due to the woolly hairs on their undersides. In spring, bitterbrush produces fragrant flowers along its branches in such abundance that the entire plant turns yellow. It can have an erect or decumbent habit but typically grows 2 to 6 ft. tall and wide, sometimes more. This is a tough and beautiful shrub for gardens in hot, dry areas.

Bitterbrush is a drought-tolerant flowering shrub that produces profusions of spring flowers.

Rhododendron macrophyllum • Ericaceae

Pacific rhododendron

HABITAT/RANGE Coastal areas, mountainous areas, forests, and thickets. Southern British Columbia to central California at low to mid elevations. Grows in and west of the Cascade Mountains. Washington state flower.

SEASONAL INTEREST Pale to deep pink flowers in spring to midsummer depending on location. Evergreen.

WILDLIFE VALUE Attracts bees, bumble bees, butterflies, hummingbirds, and other pollinators. Butterfly and moth host plant. Seeds eaten by birds. Provides cover. Browsed by mountain beavers. Not appropriate for honey bee gardens as species of *Rhododendron* contain toxins that become concentrated in honey.

CULTIVATION Part shade and moist to seasonally moist, well-drained, slightly acidic, humus-rich soil. Shade loving but tolerates sun, especially in coastal areas. Suffers leaf burn in intense sunlight. Water to establish and continue to provide supplemental water as needed. Drought tolerant once established if properly sited. Can grow under conifers. Great for shady hedges, screens, and forest edges. Best grown west of the Cascades. Has mycorrhizal relationships. Mulch.

Rhododendrons flourishing in Pacific Northwest gardens is a common and iconic sight, yet there are surprisingly few species of *Rhododendron* native to the region. Among them, *R. macrophyllum* can truly rival its cultivated cousins. This evergreen shrub has large leathery leaves and huge, pale to deep pink flowers. It grows 5 to 10 ft. tall and wide, but in deep shade can grow up to 25 ft. tall becoming treelike. Many cultivars are sold in nurseries and some exotic species may be mislabeled as native. Choose plants wisely. Never dig from the wild; legally protected in some areas.

Pacific rhododendron is a shade-loving evergreen with huge pink flowers that attract bumble bees.

Rhus aromatica • Anacardiaceae

Lemonade sumac

HABITAT/RANGE Streambanks, canyons, hillsides, thickets, open woodlands, and disturbed areas. Widely distributed in the United States. Occurs chiefly east of the Rocky Mountains and west to southern Idaho, Oregon, and California. Low to mid elevations, higher in some places. Grows east of the Cascade Mountains.

SEASONAL INTEREST Yellowish flowers in spring. Orange-red fruits summer to fall. Fall color. Deciduous.

WILDLIFE VALUE Attracts bees, butterflies, and other pollinators. Butterfly and moth host plant. Fruits eaten by birds and small animals. Provides good cover and nesting sites. Browsed by deer and elk when more palatable forage is scarce.

CULTIVATION Sun to light shade and well-drained soil. Prefers good drainage but grows in a variety of soil types including clay. Water to establish. Drought tolerant once established. Can form thickets by layering stems. Good for hedgerows, bird habitat, and erosion control. Useful for bank stabilization but not tolerant of flooding. Female plants produce fruits. Can take heavy pruning. Mulch.

Lemonade sumac has many outstanding qualities, foremost of which is the vibrant fall colors its fragrant, deeply lobed leaves often display. This deciduous shrub also produces showy clusters of orange-red fruits that may persist into winter. When fresh, these slightly hairy fruits can be soaked in cold water to produce a lemonade-like beverage. The creamy yellow flowers are small and bloom in clusters before the leaves emerge. Usually growing wider than tall, it reaches as much as 6 ft. tall and 8 ft. wide. Also referred to as *Rhus trilobata*.

Lemonade sumac produces fruits that feed birds and attractive foliage that turns beautiful colors in fall.

Rhus glabra • Anacardiaceae

Smooth sumac

HABITAT/RANGE Dry, open areas, hillsides, roadsides, canyons, and draws. Occurs throughout the United States, southern Canada, and northern Mexico at low to mid elevations. Does not naturally occur in California. Grows east of the Cascade Mountains.

SEASONAL INTEREST Spikes of greenish yellow flowers in midspring to midsummer. Red fruits in summer to fall, often persisting into winter. Vivid fall color. Deciduous.

WILDLIFE VALUE Attracts bees, butterflies, and other pollinators. Butterfly host plant. Fruits eaten by birds and other wildlife. Provides cover and nesting sites. Browsed by deer when more palatable forage is scarce.

CULTIVATION Full sun and moist to dry, well-drained soil. Water to establish. Drought tolerant once established. Quick growing and rhizomatous. Can spread to form thickets; not appropriate for small spaces. Dry soils slow spreading. Female plants produce fruits. Good for bird habitat and erosion control. Mulch.

Smooth sumac is well-known for its brilliant displays of fall foliage. The large, pinnately compound leaves give the plant a tropical look and turn brilliant shades of scarlet red in autumn. This display of color is amplified by showy, cone-shaped clusters of hairy, red fruits. The fruits are long lasting and can be soaked in cold water to make a lemonade-like beverage when fresh. The greenish yellow flowers are small, but bloom in large terminal clusters. Growing 5 to 15 ft. tall and wide in seemingly harsh habitats, this deciduous shrub to small tree can spread by woody rhizomes to form thickets, which creates ample wildlife habitat but is too aggressive for formal gardens.

Smooth sumac produces cone-shaped clusters of deep red fruits.

Ribes aureum var. *aureum* • Grossulariaceae

Golden currant

HABITAT/RANGE Seasonally moist sites in sagebrush-steppe, openings in pine forests, and along streams. Washington to California and east to the Rocky Mountains at low to high elevations. Grows east of the Cascade Mountains.

SEASONAL INTEREST Yellow flowers in spring. Edible fruits early summer. Fall color. Deciduous.

WILDLIFE VALUE Attracts bees, bumble bees, butterflies, hummingbirds, and other pollinators. Specialist bee host. Attracts beneficial insects. Butterfly and moth host plant. Fruits eaten by birds and wildlife. Provides cover.

CULTIVATION Sun to light shade and moist to seasonally dry, well-drained soil. Water to establish. Drought tolerant once established. Afternoon shade and occasional supplemental water is beneficial in arid areas. Suckers but is easily controlled. Great for pollinator plantings, edible landscapes, and xeric gardens, especially east of the Cascades. Mulch. *Ribes* are alternate hosts for white pine blister rust; do not plant in areas where five-needled pines are present.

Golden currant is an incredible plant with almost year-round interest. It quickly grows 3 to 10 ft. tall and wide, producing copious spring blooms, delicious summer fruits, and colorful fall foliage. When in bloom this showy shrub is a magnet for hummingbirds, bumble bees, and butterflies. The yellow flowers are incredibly fragrant, emitting a pleasant clovelike scent. Both deciduous and delicious, it yields clusters of berries that turn orange to deep red when ripe, beckoning birds who relish the fruits. The berries are also edible to humans and choice! Be aware that *Ribes aureum* var. *villosum* (previously *R. odoratum*) is native east of the Rocky Mountains.

Golden currant is a prolific bloomer with delicious fruits.

Ribes divaricatum • Grossulariaceae

Coast black gooseberry

HABITAT/RANGE Moist, open forests and hillsides from British Columbia to California. Low elevations. Grows mainly west of the Cascade Mountains.

SEASONAL INTEREST Fuchsialike flowers in spring. Purplish black fruits in summer. Some fall color. Deciduous.

WILDLIFE VALUE Attracts bees, butterflies, hummingbirds, and other pollinators. Attracts beneficial insects. Butterfly host plant. Specialist bee host. Fruits eaten by birds and other wildlife. Provides cover.

CULTIVATION Sun to part shade and moist, humus-rich soil. Water to establish. Drought tolerant once established if properly sited. Thorny, place away from pathways. Good for hedgerows, woodland edges, and coastal gardens. Prune after flowering if needed. Mulch. *Ribes* are alternate hosts for white pine blister rust; do not plant in areas where five-needled pines are present.

Coast black gooseberry grows 3 to 6 ft. tall and wide with a pleasant arching habit and branches that bear attractively lobed leaves and thorns at their nodes. The pendulous flowers are reminiscent of fuchsias and attract an array of pollinators. They yield smooth, purplish black berries that are edible to birds as well as humans. This deciduous shrub prefers moist soils but is surprisingly drought tolerant and grows well in low-water landscapes. A few variations occupy different parts of the species' range; use locally and ethically sourced plant material. Sadly, one variation in Southern California is likely extinct due to urbanization and development. Many species of *Ribes* are native to the region. Swamp gooseberry (*R. lacustre*) also produces black berries and grows on both sides of the Cascades.

Ribes divaricatum has small, fuchsialike flowers and a pleasant arching habit.

Ribes sanguineum • Grossulariaceae

Red flowering currant

HABITAT/RANGE Open to wooded slopes and valleys from British Columbia to California. Disjunct and imperiled in Idaho. Low to mid elevations. Grows mainly west of the Cascade Mountains.

SEASONAL INTEREST Light to deep pink, occasionally white, flowers in spring to early summer. Blue-black berries in summer. Fall color. Deciduous.

WILDLIFE VALUE Early-season food source for bees, bumble bees, butterflies, hummingbirds, and other pollinators. Specialist bee host. Attracts beneficial insects. Butterfly and moth host plant. Fruits eaten by birds and other wildlife.

CULTIVATION Sun to part shade and moist to seasonally dry, well-drained soil. Water to establish. Drought tolerant once established if properly sited. Prefers some shade and occasional supplemental water in hot, dry areas. A beautiful plant for pollinator-friendly gardens. Can be grown as a hedge. Prune after flowering if needed. Mulch. *Ribes* are alternate hosts for white pine blister rust; do not plant in areas where five-needled pines are present.

Red flowering currant is a popular landscaping plant and for good reason. The light to deep pink, sometimes white, flowers of this deciduous shrub are plentiful and very showy. These flowers are highly attractive to hummingbirds that rely on the early-season nectar as they migrate. It yields bluish black berries with a whitish cast that are edible to wildlife and humans. Best described as insipid, they are high in pectin and can be used in jams. Red flowering currant has an erect, open habit and grows quickly, reaching anywhere from 3 to 10 ft. tall and wide. There are many cultivars of this beloved shrub; choose plants wisely and favor planting true natives.

Red flowering currant has vibrantly colored flowers that attract hummingbirds.

Rosa nutkana • Rosaceae

Nootka rose

HABITAT/RANGE Coastal areas, riparian areas, open forests and woodlands, thickets, rocky slopes, and meadows. Alaska to California and east to the Rocky Mountains at low to mid elevations. Grows on both sides of the Cascade Mountains.

SEASONAL INTEREST Light to deep pink flowers in late spring to midsummer depending on location. Fruits ripen early fall and persist into winter. Some fall color. Deciduous.

WILDLIFE VALUE Attracts bees, bumble bees, and other pollinators. Butterfly host plant. Attracts beneficial insects. Fruits eaten by wildlife. Creates excellent cover and nesting sites. Browsed by deer, elk, and other wildlife.

CULTIVATION Sun to part shade and wet to seasonally dry, humus-rich soil. Adaptable to a wide range of conditions. Water to establish. Drought tolerant once established if properly sited. Rhizomatous; spreads to form thickets. Drier soils slow spreading. Typically thorny; place away from pathways. Great for pollinator and rain gardens, hedgerows, and food forests. Mulch.

Nootka rose has the largest flowers and fruits of the region's native roses. The pink flowers are divinely fragrant and feed pollinators like mason bees. Growing 3 to 6 ft. tall and wide or more, it can spread to form thorny thickets creating excellent bird habitat. The fruits, called hips, persist on the plants through the winter providing lasting interest in the garden and valuable winter food for wildlife. The hips are high in vitamin C and can be eaten fresh or dried after removing the seeds. This is an adaptable and easy-to-grow deciduous shrub with multiple benefits for gardeners and wildlife.

Nootka rose has large, fragrant flowers and red fruits called hips.

Rosa woodsii • Rosaceae

Pearhip rose

HABITAT/RANGE Riparian areas, open forests, thickets, hillsides, and shrub-steppe. Alaska to California and east beyond the Rocky Mountains. Low to mid elevations, sometimes higher. Grows mainly east of the Cascade Mountains.

SEASONAL INTEREST Light to deep pink flowers in late spring to summer. Fruits ripen early fall and persist into winter. Some fall color. Deciduous.

WILDLIFE VALUE Attracts bees, bumble bees, and other pollinators. Butterfly host plant. Attracts beneficial insects. Fruits eaten by wildlife. Creates excellent cover and nesting sites. Browsed by deer, elk, and other wildlife.

CULTIVATION Sun to part shade and moist to seasonally dry soil. Adaptable to a wide range of soil types. Water to establish. Drought tolerant once established. Rhizomatous; spreads to form thickets. Typically thorny; place away from pathways. Great for pollinator gardens, hedgerows, and edible landscapes. Useful for erosion control and wildlife habitat. Mulch.

Pearhip rose is a good choice for gardeners east of the Cascades. It bears clusters of flowers that are wonderfully fragrant and popular with pollinators. The hips ripen in late fall and persist into winter, providing long-lasting interest and food for wildlife. High in vitamin C, they can be eaten after removing the seeds. This deciduous shrub grows quickly, reaching 3 to 6 ft. tall, and continues to spread by rhizomes. A widespread and variable species; use locally sourced and adapted plants.

Rosa woodsii is a drought-tolerant shrub with fragrant flowers and edible fruits.

Rubus leucodermis • Rosaceae

Blackcap raspberry

HABITAT/RANGE Forest edges and openings, thickets, fields, and disturbed areas. Southern British Columbia to Southern California and east to the Rocky Mountains at low to mid elevations. Grows on both sides of the Cascade Mountains.

SEASONAL INTEREST White flowers in spring to summer depending on location. Edible fruits in summer. Deciduous.

WILDLIFE VALUE Attracts bees, bumble bees, butterflies, and other pollinators. Moth host plant. Attracts beneficial insects. Fruits eaten by wildlife. Provides cover and nesting sites. Old canes used by cavity-nesting bees. Browsed by various mammals.

CULTIVATION Sun to part shade and moist to seasonally dry, well-drained soil. Tolerates a variety of soil types. Water to establish. Drought tolerant once established if properly sited. Canes are typically biennial, producing flowers and fruits the second year. Cut out spent canes, leaving 1 ft. or more as nesting habitat for bees. Not rhizomatous, but branches can root where they touch the ground forming new plants. Plant in edible landscapes, farms, hedgerows, and pollinator habitat. Mulch.

Blackcap raspberry is cultivated for its delicious edible fruits, which ripen to a dark purple. They are high in antioxidants and have been used to make dyes. The fruits develop from white flowers that are popular with bumble bees. Also called whitebark raspberry because of the white to bluish color of its well-armed, thorny branches, this deciduous shrub has an arching habit and grows up to 6 ft. tall and wide. Red raspberry (*Rubus idaeus*) produces red fruits and grows east of the Cascades.

Blackcap raspberry fruits are edible and choice.

Rubus nutkanus/Rubus parviflorus • Rosaceae

Thimbleberry

HABITAT/RANGE Widely distributed in a variety of habitats. Alaska to northern Mexico and east to the Rocky Mountains and Great Lakes at low to high elevations. Grows on both sides of the Cascade Mountains.

SEASONAL INTEREST White flowers in spring to summer depending on location. Edible fruits in summer. Some fall color. Deciduous.

WILDLIFE VALUE Attracts bees, butterflies, and other pollinators. Attracts beneficial insects. Moth host plant. Fruits eaten by many species. Provides cover and nesting sites for bees and wildlife. Browsed by deer, elk, and other animals.

CULTIVATION Shade to sun and moist to seasonally dry, humus-rich, well-drained soil. Water to establish and continue to provide supplemental water as needed. Tolerates seasonally dry conditions. Prefers cool, moist sites. Spreads vigorously. Canes are typically biennial. Thin spent canes or leave as nesting sites for bees. Plant along forest edges, shady hillsides, riparian areas, and in edible landscapes. Useful for erosion control. Mulch.

Thimbleberry has multiple benefits for gardeners, though its vigorous growth makes it most appropriate for wilder areas of the landscape. With an erect habit, this thornless deciduous shrub grows 3 to 5 ft. tall, sometimes taller, and indefinitely wide as it continues to spread. The large, soft leaves create a dense, attractive cover while the showy white flowers attract pollinators. The edible red fruits are delicious, high in vitamin C, and best eaten as soon as they are picked.

Thimbleberry creates a lush cover and produces edible fruits.

Rubus spectabilis • Rosaceae

Salmonberry

HABITAT/RANGE Moist forests, wetlands, coastal areas, and riparian areas. Alaska to Northern California at low to mid elevations. Grows in and west of the Cascade Mountains. Disjunct populations in Idaho listed as rare.

SEASONAL INTEREST Purplish pink flowers in spring to midsummer depending on location. Edible orange to red fruits. Some fall color. Deciduous.

WILDLIFE VALUE Attracts hummingbirds, bees, bumble bees, butterflies, and other pollinators. Attracts beneficial insects. Moth host plant. Fruits eaten by wildlife. Provides cover, nesting sites, and nesting materials for birds and cavity-nesting bees. Browsed by deer, elk, and other wildlife.

CULTIVATION Shade to sun and moist to wet, humus-rich soil. Water to establish and continue to provide supplemental water as needed. Prefers cool, moist, shady sites. Rhizomatous; spreads vigorously in ideal conditions forming thickets. Drier soils limit spreading. Plant in riparian areas, rain gardens, food forests, and along forest edges. Best suited for gardens west of the Cascades. Mulch.

Salmonberry is a lovely deciduous understory shrub that feeds pollinators, wildlife, and people. It thrives in moist maritime climates and cool, shady mountain forests. It is fast growing with an erect to arching habit and typically thorny branches. Reaching 3 to 10 ft. tall and wide, it spreads by woody rhizomes to form thickets that provide excellent nesting habitat for birds and cavity-nesting bees. The flowers are a deep purplish pink and an important early-season food source for pollinators. Large, colorful orange to red fruits make it a great addition to edible landscapes.

Salmonberry is a shade-loving shrub with deep purplish pink flowers and edible fruits.

Sambucus racemosa • Adoxaceae/Viburnaceae

Red elderberry

HABITAT/RANGE Coastal areas, riparian areas, forests, and wet meadows. Alaska to California and east across North America at low to high elevations. Circumboreal. Grows on both sides of the Cascade Mountains.

SEASONAL INTEREST White flowers early spring to midsummer depending on location. Red fruits in summer. Deciduous.

WILDLIFE VALUE Attracts bees, butterflies, hummingbirds, and other pollinators. Butterfly and moth host plant. Attracts beneficial insects. Fruits eaten by wildlife. Provides cover and nesting sites. Pithy stems used by cavity-nesting bees like mason bees. Browsed by deer and elk.

CULTIVATION Shade to sun and moist, humus-rich soil. Water to establish and continue to provide supplemental water as needed. Tolerates seasonally dry conditions once established if properly sited. When pruning leave at least 4 to 6 in. above the leaf node to create nesting sites for bees. Plant in woodland gardens, riparian areas, coastal gardens, hedgerows, bee gardens, and backyard bird habitat. Mulch.

Red elderberry is a deciduous shrub with attractive flowers, fruits, and leaves. Growing 6 to 20 ft. tall and wide, it can be a multibranched shrub or small tree. It has pyramidal clusters of creamy white flowers and pinnately compound leaves. Its showy, bright red fruits are an important food for birds, but all parts are the plant are considered toxic to humans. There are a few, somewhat geographically divided, variations. Coast red elderberry (*Sambucus racemosa* var. *arborescens*) is common in and west of the Cascades. Black elderberry (*S. racemosa* var. *melanocarpa*) occurs in and east of the Cascades and has purplish black fruits.

Red elderberry is showy in bloom but even showier in fruit.

Spiraea douglasii • Rosaceae

Douglas's spiraea

HABITAT/RANGE Moist sites, wet meadows, riparian areas, coastal areas, and wetlands. Alaska to California and east to Montana at low to high elevations. Grows on both sides of the Cascade Mountains.

SEASONAL INTEREST Pink flowers in summer. Fall color. Deciduous.

WILDLIFE VALUE Attracts bees, bumble bees, butterflies, hummingbirds, and other pollinators. Attracts beneficial insects. Butterfly and moth host plant. Seeds eaten by birds and wildlife. Provides cover and nesting sites. Browsed by deer.

CULTIVATION Sun to part shade and moist to wet soil. Grows in a variety of soil types including clay. Water to establish and continue to provide supplemental water as needed. Plant in moist areas such as lake margins, riparian areas, rain gardens, and irrigated gardens. Tolerates flooding. Rhizomatous; spreads to form thickets. Excellent for wetland restoration and hedgerows. Prune as needed. Mulch.

Douglas's spiraea is a beautiful deciduous shrub with long, dense clusters of showy, bright pink flowers that are a magnet for bumble bees and butterflies. Growing up to 6 ft. tall, this moisture-loving shrub spreads vigorously by rhizomes to form thickets that provide excellent habitat for songbirds such as red-winged blackbirds, as well as waterfowl. It is a good species for riparian bank stabilization as it tolerates flooding. There are a few species of *Spiraea* native to the region and they hybridize with each other. Pyramidal spiraea (*S. ×pyramidata*) is a natural hybrid between *S. douglasii* and *S. lucida*. Subalpine spiraea (*S. splendens*) grows at mid to high elevations.

Douglas's spiraea is a moisture-loving shrub with bright pink flowers loved by bumble bees.

Spiraea lucida • Rosaceae

Shinyleaf spiraea

HABITAT/RANGE Open woodlands, rocky slopes, and streambanks. British Columbia to Oregon and east beyond the Rocky Mountains at low to high elevations. Grows on both sides of the Cascade Mountains.

SEASONAL INTEREST White flowers in late spring to summer relative to elevation. May bloom sporadically into fall. Fall color. Deciduous.

WILDLIFE VALUE Attracts bees, butterflies, and other pollinators. Attracts beneficial insects. Butterfly larval host. Provides cover for small animals. Usually ignored by deer.

CULTIVATION Part shade to sun and moist to seasonally dry, well-drained soil. Water to establish. Drought tolerant once established if properly sited. Prefers some shade and benefits from occasional summer water in hot, dry sites. Watering may also prolong blooming. Rhizomatous; spreads to form patches and can creep around the garden but is not aggressive. Drier soils limit spreading. Great low-growing shrub for borders, forest edges, parking strips, and semi-shady, low-water landscapes. Can be grown in containers. Mulch.

Shinyleaf spiraea is more drought and shade tolerant than Douglas's spiraea. It is a low-growing deciduous shrub, usually only reaching 2 ft. tall with a mostly erect habit and attractive leaves that provide nice fall color. Its flat-topped clusters of white flowers have a fuzzy appearance due to their long, silky stamens. *Spiraea lucida* hybridizes readily with other species of *Spiraea*, also called meadowsweets. Previously considered a variation of the Asian species *S. betulifolia* (birchleaf spiraea), many still reference it as *S. betulifolia* var. *lucida*. Be careful not to buy exotic species mistakenly marked as natives.

Spiraea lucida has lovely flat-topped clusters of fuzzy, white flowers.

Symphoricarpos albus • Caprifoliaceae

Snowberry

HABITAT/RANGE Widespread and found in a variety of habitats. Alaska to California and east to Montana. One variation grows mainly east of the Rocky Mountains. Low to mid elevations. Grows on both sides of the Cascade Mountains.

SEASONAL INTEREST Pink to white flowers in late spring to late summer. White fruits late summer through winter. Some fall color. Deciduous.

WILDLIFE VALUE Attracts bees, bumble bees, butterflies, hummingbirds, and other pollinators. Attracts beneficial insects. Larval host for checkerspot butterflies and hummingbird clearwing moths. Fruits eaten by birds and small mammals. Provides cover and nesting sites. Browsed by deer and other wildlife.

CULTIVATION Adaptable. Sun to part shade and moist to seasonally dry soil. Grows in a variety of soil types. Tolerates poor soils but prefers humus-rich soils with good drainage. Water to establish. Drought tolerant once established. Grows more robustly with some shade and moisture, more compactly in sun. Spreads vigorously by rhizomes. Good for erosion control. Makes an attractive hedge. Prune as needed. Mulch.

Bees are drawn to the copious nectar snowberry's small, pink to white flowers produce. Their pollination services help to create showy, white, berrylike fruits that give this plant its name and ornamental value. The fruits, which are not considered edible to humans, persist through winter providing lasting interest and a valuable food source for birds. This deciduous shrub grows up to 6 ft. tall and spreads vigorously. One of the region's most adaptable shrubs, it can grow in a variety of climates and conditions.

Snowberry feeds pollinators and wildlife such as ruffed grouse.

Symphoricarpos mollis • Caprifoliaceae

Creeping snowberry

HABITAT/RANGE Forest edges, thickets, and slopes from southwestern British Columbia to California. Disjunct in Idaho. Low to mid elevations. Grows on both sides of the Cascade Mountains.

SEASONAL INTEREST Pink flowers late spring to midsummer relative to elevation. White fruits late summer into winter. Deciduous.

WILDLIFE VALUE Attracts bees, bumble bees, hummingbirds, and other pollinators. Attracts beneficial insects. Butterfly and moth host plant. Fruits eaten by birds and other wildlife. Provides cover for ground-nesting birds.

CULTIVATION Sun to shade and moist to seasonally dry, well-drained soil. Water to establish. Drought tolerant once established if properly sited. Some supplemental water in sunny, dry sites will keep plants looking lush. Trailing branches root where they touch the ground forming new plants. Cut back as needed. A great pollinator-friendly groundcover. Useful for erosion control. Mulch.

Creeping snowberry is an attractive and very ornamental groundcover. This trailing deciduous shrub only grows about 1 ft. tall and has long, leafy branches that arc and creep along the ground. The small pink flowers are bell-shaped, loved by bumble bees, and produce a profusion of showy, snow-white fruits. The berrylike fruits, which are not considered edible to humans, persist on the plants long into winter. Its tolerance to various soils and conditions makes it a good candidate for a wide variety of gardens. Some nurseries may refer to this plant as *Symphoricarpos hesperius*.

Creeping snowberry is a unique trailing shrub that bears copious amounts of white, berrylike fruits, making it a highly ornamental groundcover.

Vaccinium ovatum • Ericaceae

Evergreen huckleberry

HABITAT/RANGE Coastal areas, forests, thickets, and clearings. Southwestern British Columbia to Southern California at low elevations. Grows west of the Cascade Mountains, mainly along the coast.

SEASONAL INTEREST Pink to white flowers early spring to summer. Edible purplish black berries. Evergreen.

WILDLIFE VALUE Attracts bumble bees, butterflies, hummingbirds, and other pollinators. Butterfly and moth host plant. Fruits eaten by birds and wildlife. Provides evergreen cover and nesting sites. Browsed by elk, occasionally deer.

CULTIVATION Shade to sun and moist, humus-rich, well-drained soil. Prefers acidic soils. Can grow in sandy soils. Water to establish. Drought tolerant once established if properly sited, especially in coastal gardens, otherwise provide supplemental water as needed. Will grow in full sun with adequate moisture but prefers some shade. Slow growing. Plant in coastal gardens, food forests, partly shaded gardens, or along forest edges. Great evergreen for hedges or screens. Mulch.

Evergreen huckleberry creates a beautiful hedge adorned with edible fruits. The leaves of this evergreen shrub are dark green and glossy with new growth emerging in lovely shades of peach to red. Its small, bell-shaped flowers are white to pink and popular with pollinators, while both birds and humans relish its purplish black berries. Slowly growing 3 to 12 ft. tall and 3 to 10 ft. wide depending on site conditions, this coastal shrub does best in mild maritime climates and makes a handsome addition to gardens west of the Cascades. Many species of huckleberry and wild blueberry are native to the region; this is one of the easiest to cultivate.

Vaccinium ovatum has delicious berries and makes a beautiful hedge.

Vaccinium parvifolium • Ericaceae

Red huckleberry

HABITAT/RANGE Moist forests, often on rotting logs or stumps, coastal areas, and riparian areas. Alaska to California at low to mid elevations. Grows primarily west of the Cascade Mountains.

SEASONAL INTEREST Yellowish pink flowers in spring to early summer. Edible red fruits in summer. Lacy stems and foliage. Deciduous.

WILDLIFE VALUE Attracts bees, butterflies, hummingbirds, and other pollinators. Butterfly host plant. Fruits eaten by birds and mammals. Browsed by mountain beavers, elk, and deer.

CULTIVATION Full to part shade and moist, acidic, humus-rich soil. Water to establish and continue to provide supplemental water as needed. Tolerates seasonally dry conditions once established if properly sited. A good plant for stumperies as it likes to grow on rotting wood; plant in or near rotting stumps, start with young plants, and maintain even soil moisture while establishing. Does not transplant well. Slow growing. Plant in moist woodland gardens, food forests, rain gardens, and shady riparian areas. Mulch.

Red huckleberry is a lacy, shade-loving shrub that slowly grows 3 to 12 ft. high and 2 to 6 ft. wide with an erect habit. Deciduous though sometimes semievergreen, its green, angular branches are attractive with or without their small leaves and provide structural interest. The yellowish pink flowers are pendulous and inconspicuous but feed a variety of pollinators. This graceful shrub produces tart, edible, bright red berries that are somewhat translucent and, along with the bright green foliage, look dazzling when lit up by sunbeams in the dark forest understory.

Red huckleberry has an attractive lacy look and produces edible red fruits.

Viburnum edule • Adoxaceae/Viburnaceae

Highbush cranberry

HABITAT/RANGE Moist woodlands, thickets, riparian areas, and wetlands. Widespread. Alaska to Oregon and east to Newfoundland at low to mid elevations. Grows on both sides of the Cascade Mountains.

SEASONAL INTEREST White flowers late spring to midsummer depending on location. Edible red to orange fruits. Gorgeous fall color. Deciduous.

WILDLIFE VALUE Attracts bees, butterflies, and other pollinators. Attracts beneficial insects. Moth host plant. Fruits eaten by birds and mammals. Provides cover. Browsed by elk, deer, beaver, and other wildlife.

CULTIVATION Sun to part shade and moist, humus-rich soil. Prefers well-drained soil but can grow in clay. Water to establish and continue to provide supplemental water as needed. Will grow in full sun with adequate moisture. Great for riparian areas, irrigated gardens, hedgerows, edible landscapes, and backyard bird habitat. Mulch.

Viburnum edule is an attractive deciduous shrub that puts on a lovely display of fall color with its broad, shallowly lobed leaves. Its flat-topped clusters of white flowers provide interest in spring and early summer while the red to orange, edible, cranberry-like fruits provide additional color into fall. The fruits are relished by birds, making this is an excellent plant for backyard bird habitat. Viburnums are popular landscaping plants and there are many cultivated varieties from around the world. *Viburnum edule* can be confused with *V. opulus*, also called highbush cranberry, which has two variations, one native to the northern part of the region and one native to Europe that has escaped cultivation.

Highbush cranberry is a moisture-loving shrub with edible fruits and beautiful fall color.

Viburnum ellipticum • Adoxaceae/Viburnaceae

Oval-leaved viburnum

HABITAT/RANGE Open woodlands and thickets. Washington to California at low to mid elevations. Grows mainly in and west of the Cascade Mountains. Rare in California.

SEASONAL INTEREST White flowers spring to early summer. Red to black fruits. Fall color. Deciduous.

WILDLIFE VALUE Attracts bees, butterflies, and other pollinators. Attracts beneficial insects. Moth host plant. Fruits eaten by birds. Provides cover.

CULTIVATION Sun to part shade and moist to seasonally dry, well-drained soil. Water to establish. Drought tolerant once established if properly sited. Benefits from afternoon shade and some supplemental water in hot, dry areas. Plant in hedgerows and along woodland edges. Makes a lovely specimen in ornamental gardens. Mulch.

Oval-leaved viburnum is a handsome shrub loved by gardeners and landscapers for its "three season interest." In spring it bears showy flat-topped clusters of white flowers that attract a wide variety of pollinators and beneficial insects. In summer the red, berrylike fruits turn shiny black when ripe and provide tasty snacks for birds. And in fall the oval, coarsely toothed leaves of this deciduous shrub turn bronzy shades of red. Growing 3 to 10 ft. tall and 4 to 8 ft. wide with an erect habit, it is more drought tolerant than *Viburnum edule*. Although best suited for gardens west of the Cascades, it can take a fair amount of sun and heat as long as it has adequate moisture.

Viburnum ellipticum has showy flowers that attract a variety of pollinators.

 # Trees

Abies grandis • Pinaceae

Grand fir

HABITAT/RANGE Coniferous forests in coastal and mountainous areas from southern British Columbia to Northern California and east to Montana. Low to mid elevations. Grows on both sides of the Cascade Mountains.

SEASONAL INTEREST Large upright cones. Evergreen.

WILDLIFE VALUE Wind pollinated. Attracts beneficial insects. Seeds eaten by birds and small mammals. Needles eaten by grouse. Butterfly and moth larval host. Provides food, cover, and nesting sites for birds and other wildlife. Needles provide winter browse for deer and elk, but it is not preferred forage.

CULTIVATION Sun to part shade and moist to seasonally dry, slightly acidic, humus-rich, well-drained soil. Grows in a variety of soil types. Water to establish. Drought tolerant once established if properly sited. Provide supplemental water in areas that receive less than 25 in. of rain a year. Prefers cool locations such as north-facing slopes. Mulch.

Grand fir is a beautiful evergreen that grows straight and tall, eventually well over 100 ft. with lower branches reaching 20 ft. wide or more. It is popular for its pyramidal shape along with the citrus scent emitted by its needles, which alternate in length and lay flat. The foliage is especially attractive in spring and early summer when bright green new growth accents the tips of the branches. Grand fir is more common at low elevations than most other true firs in the region. A gregarious tree, it prefers to mingle in mixed conifer forests where it adds to the diversity of shelter and forage available to wildlife.

Grand fir needles have a lovely citrusy scent.

Abies lasiocarpa • Pinaceae

Subalpine fir

HABITAT/RANGE Conifer forests in mountainous areas and subalpine slopes from Alaska to southern Oregon and through the Rocky Mountains to areas of the southwestern United States. Mid to high elevations. Grows in and on both sides of the Cascade Mountains.

SEASONAL INTEREST Purple cones. Evergreen.

WILDLIFE VALUE Wind pollinated. Attracts beneficial insects. Seeds eaten by birds and small mammals. Needles eaten by grouse and other wildlife. Butterfly and moth host plant. Provides food, cover, and nesting sites. Supports lichens that feed caribou. Not preferred browse for deer and elk.

CULTIVATION Sun to part shade and moist, well-drained soil. Water to establish. Tolerates seasonally dry conditions once established if properly sited. Prefers a cool, moist site. Able to reproduce vegetatively by layering. A good specimen for rock gardens, formal landscapes, and containers. Highly flammable, place away from buildings. Mulch.

Subalpine fir grows at high elevations in mountainous climates where deep winter snow has shaped it into a slender, spirelike evergreen. An attractive ornamental, its sometimes stunted and windswept figure has inspired its use in bonsai. Somewhat colorful for a conifer, the upturned needles are blue green, the erect cones are purple, and the bark is gray and smooth with resin blisters. This beautiful evergreen can grow up to 100 ft. tall and 15 ft. wide but is slow growing and tends to be shorter in cultivation. *Abies lasiocarpa* has borne the brunt of wild collection, which damages the fragile environments in which it grows. Make sure plants are nursery grown and ethically sourced.

Subalpine fir has an attractive, slender form.

Acer circinatum • Sapindaceae

Vine maple

HABITAT/RANGE Moist forests, riparian areas, open slopes, and clearings. British Columbia to Northern California at low to mid elevations. Grows from the coast to the eastern foothills of the Cascade Mountains.

SEASONAL INTEREST Flowers in spring. Brilliant fall color. Attractive structure. Deciduous.

WILDLIFE VALUE Attracts bees, butterflies, and other pollinators. Larval host for butterflies and moths including the western tiger swallowtail butterfly. Attracts beneficial insects. Seeds eaten by birds and mammals. Provides cover and nesting sites. Shades streams and helps create healthy fish habitat. Browsed by deer, elk, and other animals.

CULTIVATION Shade to sun and moist to seasonally moist, humus-rich, well-drained soil. Prefers moist shade. Water to establish. Tolerates seasonally dry conditions once established if properly sited. Fall color more dramatic with a little sun and will grow in full sun with adequate moisture but can suffer leaf burn. A beautiful tree for formal landscapes and woodland gardens. Use instead of Japanese maple. Mulch.

Vine maple is one of the region's most attractive trees. It is small in stature, sometimes shrublike, growing only 10 to 25 ft. tall and 5 to 20 ft. wide, with smooth, colorful branches and an elegant structure. The leaves have seven to nine pointed lobes and put on dazzling displays of red, orange, and yellow in fall. Small flowers bloom in clusters as new leaves emerge in spring. Bringing beauty and biodiversity together, it is gorgeous in the garden while supporting many native species of insects and wildlife.

Vine maple leaves turn brilliant colors in fall.

Acer glabrum • Sapindaceae

Rocky Mountain maple

HABITAT/RANGE Forests, forest edges, streambanks, and rocky slopes and ridges from Alaska to California and east to the Rocky Mountains. Low to high elevations depending on the variation. Grows on both sides of the Cascade Mountains.

SEASONAL INTEREST Flowers spring to early summer depending on location. Brilliant fall color. Deciduous.

WILDLIFE VALUE Attracts bees and other pollinators. Butterfly and moth host plant. Attracts beneficial insects. Seeds eaten by birds and mammals. Provides cover and nesting sites. Browsed by deer, elk, and other animals.

CULTIVATION Sun to part shade and moist to seasonally dry, well-drained soil. Prefers adequate moisture, humus-rich soil, and some shade, but tolerates a variety of conditions. Water to establish. Drought tolerant once established if properly sited. Use as a focal point along the woodland edge. Mulch.

Rocky Mountain maple is similar to vine maple but hardier and more tolerant of sun, drought, and poor soils. The foliage differs from vine maple in the leaves having only three to five pointed lobes, but the tendency toward brilliant displays of fall color is the same. This is a small deciduous tree, sometimes shrublike, that grows up to 30 ft. tall and 20 ft. wide. The flowers bloom in hanging clusters as the new leaves emerge in spring and the winged seeds are eaten by birds and other wildlife. This versatile tree is a worthy candidate for all sorts of Northwest landscapes. The species is comprised of a few variations; use locally sourced plant material when possible.

Rocky Mountain maple is a small, drought-tolerant tree.

Acer macrophyllum • Sapindaceae

Big-leaf maple

HABITAT/RANGE Woodlands, forests, riparian areas, and coastal areas. British Columbia to California at low to mid elevations. Grows on both sides of the Cascade Mountains, though mainly west.

SEASONAL INTEREST Chartreuse flowers early spring to early summer depending on location. Large, winged seeds. Fall color. Deciduous.

WILDLIFE VALUE Early-season food source for bees, butterflies, and other pollinators. Butterfly and moth host plant. Attracts beneficial insects. Seeds eaten by birds and mammals. Provides shade, cover, and nesting sites. Supports epiphytic life. Shades streams and creates healthy fish habitat. Browsed by deer, elk, and other animals.

CULTIVATION Sun to shade and moist to seasonally dry, humus-rich, well-drained soil. Water to establish. Drought tolerant once established if properly sited. Prefers a cool, moist site. A beautiful shade tree for everything from formal landscapes to riparian plantings. Protect from deer when young. Fast growing. Mulch.

Big-leaf maple is a stately deciduous tree that grows 30 to 100 ft. tall and 20 to 60 ft. wide, providing ample and luxurious shade with its huge leaves. In moist lowland forests, communities of epiphytic life such as licorice fern and mosses flourish on its moisture-retentive, furrowed bark. Its large, hanging clusters of early-blooming chartreuse flowers are important for pollinators like mason bees. The flowers are edible and can be used in salads, their flavor sweetest when young. In autumn the large leaves light up dark forests by turning brilliant yellow and add to soil health by creating a nutrient-rich mulch when they fall.

Big-leaf maple has showy clusters of chartreuse flowers.

Alnus rubra • Betulaceae

Red alder

HABITAT/RANGE Moist slopes, flood plains, disturbed sites, riparian areas, and coastal areas. Alaska to California at low elevations. Grows mainly west of the Cascade Mountains, with disjunct populations east and in Idaho.

SEASONAL INTEREST Catkins in early spring. Fall color. Deciduous.

WILDLIFE VALUE Wind pollinated. Bees collect pollen and resin. Larval host for many butterflies and moths. Seeds eaten by birds. Provides cover and nesting sites. Shades waterways and helps create healthy fish habitat. Supports epiphytic life. Browsed by deer, elk, beavers, and other wildlife.

CULTIVATION Sun to part shade and moist, humus-rich soil. Water to establish. Tolerates seasonally dry conditions once established if properly sited. Tolerates flooding and poor soils. Grows in sandy and clay soils. Plant in riparian areas and low elevation woodlands. Mulch.

Red alder doesn't just fit a habitat, it creates it. This pioneer species is first to the scene after a disturbance in moist areas where it grows quickly, reaching 40 to 80 ft. tall and 30 to 40 ft. wide, builds soils with nitrogen-fixing bacteria, and provides shade, shelter, and forage for wildlife. It is an important butterfly larval host, and its smooth bark is home to crustose lichens. It produces both male and female catkins that often bloom before leaves emerge, the female catkins becoming hard and conelike. The leaves of this deciduous tree turn yellow in fall and enrich the soil when they drop. Better candidates for gardens east of the Cascades are white alder (*Alnus rhombifolia*) and mountain alder (*A. incana*); the latter also grows at higher elevations.

Red alder is fast growing and hosts nitrogen-fixing bacteria.

Arbutus menziesii • Ericaceae

Pacific Madrone

HABITAT/RANGE Coastal areas, forests, rocky slopes, and open areas. British Columbia to California at low to mid elevations. Grows west of the Cascade Mountains.

SEASONAL INTEREST White flowers in spring. Orange-red fruits may persist into winter. Attractive bark. Evergreen.

WILDLIFE VALUE Flowers attract hummingbirds, bees, bumble bees, butterflies, and other pollinators. Butterfly and moth larval host. Attracts beneficial insects. Fruits eaten by birds and mammals. Provides cover and nesting sites. Browsed by deer and elk.

CULTIVATION Sun to light shade and average to slightly acidic, sharply draining soil. Tricky to establish and does not transplant well. Start with young, seed-propagated plants. Only water the first year to establish. Needs dry summer conditions once established. Young plants prefer afternoon shade. Keep in mind when siting that it sheds older leaves and bark, giving it a "messy" reputation. Hardy to USDA zone 6a. A beautiful specimen tree. Protect young plants from deer. Mulch.

Pacific madrone is a stunning broadleaf evergreen. The smooth, shapely trunk and branches have mahogany to burnt-umber exfoliating bark that peels away to reveal shades of yellowish green. This tree may be tricky to establish but it is long lived and can grow quickly once it gains a foothold. The fragrant, white, urn-shaped flowers bloom in flamboyant clusters that attract pollinators like hummingbirds. The flowers become bumpy, red-orange fruits that birds love and are edible to humans. Growing anywhere from 20 to 100 ft. tall, this is a beautiful, drought-tolerant specimen for coastal gardens and sunny, dry sites west of the Cascades.

Pacific madrone is a unique tree with an elegant structure.

Calocedrus decurrens • Cupressaceae

Incense cedar

HABITAT/RANGE Dry, open areas, meadows, and mixed-conifer forests. Mount Hood in Oregon south through California to Baja California. Low to high elevations. Grows on both sides of the Cascade Mountains.

SEASONAL INTEREST Small cones late summer. Evergreen.

WILDLIFE VALUE Wind pollinated. Attracts beneficial insects. Birds glean insects from trees. Provides cover and nesting sites for a variety of wildlife. Browsed by deer.

CULTIVATION Sun to light shade and well-drained soil. Prefers seasonally moist, humus-rich soil but is adaptable to a wide range of soil types. Water to establish. Drought tolerant once established. A beautiful specimen for landscapes with enough space to accommodate its size. Useful for screening and windbreaks. Can be pruned as a hedge. Mulch.

Incense cedar is a beautiful and tough evergreen tree able to thrive in a variety of conditions. It can grow in hot, dry sites with poor soils, making it a rugged and drought-tolerant alternative to western red cedar. It is a large tree, eventually growing 50 to 100 ft. tall, sometimes taller, and more than 20 ft. wide, with a typically dense, conical shape and scalelike leaves that form flat sprays. By late summer the interesting seed cones mature, consisting of ovate, reddish brown scales with slightly convex outer scales that peel back in a way most people describe as looking like a duck's bill. This tree can be sheared as a hedge much like arborvitae or become a stately specimen in formal landscapes.

Incense cedar tolerates drought and poor soils.

Celtis reticulata • Cannabaceae

Netleaf hackberry

HABITAT/RANGE Rocky slopes, often near rivers, in arid areas from Washington to Oregon, east to Idaho and south through the Rocky Mountains to the southwestern United States, Southern California, and Mexico. Low to mid elevations. Grows east of the Cascade Mountains.

SEASONAL INTEREST Small flowers in spring. Edible purplish red fruits. Some fall color. Deciduous.

WILDLIFE VALUE Fruits eaten by wildlife, especially birds. Birds glean insects from plants. Wind pollinated. Attracts beneficial insects. Butterfly larval host. Provides shade, shelter, nesting sites, and nesting materials. Browsed by deer and other wildlife.

CULTIVATION Sun to light shade and well-drained soil. Water to establish. Drought tolerant once established. Benefits from deep, infrequent summer irrigation. Tolerates heat and poor soils. Useful as a shade tree in arid areas. Known allelopath; leaf litter may inhibit the development of nearby plants. Mulch.

Netleaf hackberry is a rugged deciduous tree that tolerates heat, drought, and poor soils but prefers some amount of moisture as it is often found near rivers and seeps in dry areas. Its inconspicuous flowers become purplish red fruits with a thin, sweet pulp that is edible to humans and beloved by birds such as flickers and waxwings. The prominently veined leaves commonly host gall-producing insects. These leaf galls do not harm the plant and provide an important food source for birds. Taking the form of a shrub or thick-trunked tree, it grows to 30 ft. tall and wide, possibly more, and is best grown east of the Cascades.

Netleaf hackberry grows in harsh habitats where it provides vital shade and shelter, as well as food for wildlife.

Cornus nuttallii • Cornaceae

Pacific dogwood

HABITAT/RANGE Forests, streambanks, clearings, and coastal areas. British Columbia to California at low to mid elevations. Disjunct and critically imperiled in Idaho. Grows mainly in and west of the Cascade Mountains. Provincial flower of British Columbia.

SEASONAL INTEREST White blooms spring to early summer, often again in fall. Red fruits. Gorgeous fall color. Deciduous.

WILDLIFE VALUE Attracts bees and other pollinators. Butterfly larval host. Attracts beneficial insects. Specialist bee host. Fruits eaten by birds and small mammals. Provides cover. Young plants browsed by deer and elk.

CULTIVATION Part shade and moist to seasonally dry, humus-rich, well-drained soil. Tolerates full sun but may suffer leaf burn. Water to establish. Prefers dry summer conditions once established. Can be finnicky and difficult to establish. A beautiful specimen for woodlands and formal gardens. Susceptible to anthracnose fungus. Mulch.

Pacific dogwood is dazzling in spring when it covers itself in huge, white, saucerlike blooms, often flowering again in fall before putting on brilliant displays of fall foliage. What look like flower petals are actually large white bracts surrounding a dense cluster of small flowers, which become showy red fruits that feed birds and other wildlife. This shade-loving deciduous tree can grow 30 to 50 ft. tall and 20 to 25 ft. wide. Nursery availability of this desirable ornamental may be limited as it is difficult to cultivate in containers. It does not transplant well and should never be dug from the wild. 'Eddie's White Wonder' is a commonly used hybrid of this species and an eastern North American native, *Cornus florida*. Favor planting true natives.

Pacific dogwood has large, saucerlike blooms.

Crataegus douglasii • Rosaceae

Black hawthorn

HABITAT/RANGE Open forests and forest edges, riparian areas, and thickets. Northern British Columbia to California and east beyond the Rocky Mountains at low to mid elevations. Grows on both sides of the Cascade Mountains.

SEASONAL INTEREST White flowers late spring to early summer. Dark burgundy to black fruits summer to fall. Fall color. Deciduous.

WILDLIFE VALUE Attracts bees, butterflies, hummingbirds, and other pollinators. Important butterfly host plant. Attracts beneficial insects. Fruits eaten by birds and mammals. Provides excellent cover and nesting sites for birds. Browsed by deer.

CULTIVATION Sun to part shade and moist to seasonally dry soil. Water to establish. Drought tolerant once established if properly sited. Vigorous and suckering, thicket forming; not appropriate for small spaces. Place in wilder parts of the landscape where it can spread. Plant in riparian areas, hedgerows, and wildlife habitat plantings. Useful for erosion control. Mulch.

Black hawthorn is a beautiful tree to large shrub with showy flowers and fruits, although its vigorous suckering habit limits its use for formal plantings. Growing 10 to 30 ft. tall and 10 to 20 ft. wide, it can be maintained as a single-trunked tree but removing suckers is bothersome. It has stiff thorns and if allowed to spread will form a thorny thicket, offering the kind of protective habitat birds love. The white flowers and dark fruits bloom and develop in clusters. The fruits are eaten by wildlife and are edible to humans as well. Many species of hawthorn are found in the region including naturalized exotics.

Crataegus douglasii has lovely clusters of white flowers that attract pollinators.

Frangula purshiana • Rhamnaceae

Cascara

HABITAT/RANGE Forest understories and edges, moist places, clearings, coastal areas, and riparian areas. British Columbia to California and east to western Montana at low to mid elevations. Grows on both sides of the Cascade Mountains.

SEASONAL INTEREST Small flowers in spring to early summer. Purplish black fruits summer to fall. Fall color. Deciduous.

WILDLIFE VALUE Attracts bees, butterflies, hummingbirds, and other pollinators. Butterfly and moth larval host. Attracts beneficial insects. Fruits relished by birds and mammals. Provides cover and nesting sites. Browsed by deer and elk.

CULTIVATION Sun to shade and moist, humus-rich, well-drained soil. Grows in a variety of soil types and light conditions. Water to establish and continue to provide supplemental water as needed. Tolerates seasonally dry conditions if properly sited. Benefits from afternoon shade in hot areas. Excellent for woodland gardens, forest edges, hedgerows, riparian areas, and bird habitat. Mulch.

Cascara has many qualities that make it desirable in gardens. Growing up to 30 ft. tall and about 20 ft. wide it takes the form of a small tree or shrub and has broad, shiny leaves with prominent veins that turn shades of yellow in fall. The inconspicuous flowers produce showy, purplish black, berrylike fruits that are a huge attraction to birds and other wildlife. Cascara has long been used by pharmaceutical companies and herbalists for its laxative effects. United Plant Savers lists it as a species at risk of being overharvested from the wild, so grow it if you want to use it. Previously classified as *Rhamnus purshiana*.

Cascara has attractive, prominently veined leaves and berrylike fruits that turn purplish black when ripe.

Fraxinus latifolia • Oleaceae

Oregon ash

HABITAT/RANGE Coastal areas, riparian areas, wetlands, forest edges, and flood plains. Southern Vancouver Island to California, generally at low elevations. Rare in British Columbia. Grows mainly in and west of the Cascade Mountains.

SEASONAL INTEREST Small greenish yellow flowers in spring. Fall color. Deciduous.

WILDLIFE VALUE Wind pollinated but still visited by bees. Butterfly larval host. Attracts beneficial insects. Seeds eaten by birds including wood ducks. Provides cover and nesting sites. Helps create healthy riparian habitat, which benefits many species. Browsed by deer, elk, and beaver.

CULTIVATION Sun to part shade and moist, humus-rich soil. Grows in a variety of soil types including clay and sandy soils. Water to establish and continue to provide supplemental water as needed. Tolerates seasonally dry conditions once established but prefers moist habitats and endures seasonal flooding. Plant in riparian areas, pond margins, and other wet spots. Susceptible to fungal blight, which deforms leaves but does not harm the plant. Mulch.

The only species of ash native to Oregon and Washington and a denizen of moist places, Oregon ash is a common sight along bodies of water west of the Cascades, particularly in its namesake state. The attractive, pinnately compound leaves are vibrant green in spring, turning shades of yellow in fall. This deciduous tree has inconspicuous male and female flowers on separate plants, with female plants producing hanging clusters of winged seeds that provide food for birds. It grows quickly, reaching 30 to 80 ft. tall and 15 to 40 ft. wide.

Oregon ash is a moisture-loving tree found in riparian and wetland habitats.

Larix occidentalis • Pinaceae

Western larch

HABITAT/RANGE Mountainous areas, forests, clearings, disturbed areas, and mountain valleys. British Columbia to central Oregon and east to Montana, mainly at mid elevations. Grows in and east of the Cascade Mountains.

SEASONAL INTEREST Fall color. Deciduous conifer.

WILDLIFE VALUE Important source of food, cover, and nesting habitat for wildlife. Wind pollinated. Seeds eaten by birds and mammals. Birds glean insects off plants. Occasionally browsed by deer and elk.

CULTIVATION Sun to light shade and well-drained soil. Water to establish. Drought tolerant once established if properly sited. Prefers sun and cool, moist spots. Naturally grows at higher elevations and is more tolerant of cold than heat. Tall and fast growing; give it plenty of head room. Uncrowded trees will maintain the best structure. Mulch.

Like a torch in fall, western larch lights up mountain slopes with brilliant shades of yellow. The annual shedding of this deciduous conifer's needles is a sublime event and provides some of our best fall color in the expanse of evergreen forests. The naked branches in winter are starkly striking against a cold blue sky, and the soft green needles, arranged spirally in tufts along the branches, always return with flare in spring. This long-lived, fire-adapted tree can grow over 200 ft. tall and has a narrow form. Also called western tamarack, it grows in cool, moist, rocky, mineral mountain soils and will need occasional supplemental water in dry, low-elevation areas.

Western larch is a deciduous conifer that turns bright yellow before shedding its needles for winter.

Picea engelmannii • Pinaceae

Engelmann spruce

HABITAT/RANGE Mountain slopes, streams, valleys, and subalpine ridges. British Columbia to Northern California and east to the Rocky Mountains mainly at mid to high elevations. Widespread in mountainous areas of western North America. Grows primarily in and east of the Cascade Mountains.

SEASONAL INTEREST Blue-green foliage. Evergreen.

WILDLIFE VALUE Wind pollinated. Attracts beneficial insects. Larval host for the pine white butterfly. Seeds eaten by birds and small mammals. Provides evergreen cover and nesting sites. Needles eaten by grouse. Occasionally browsed by deer.

CULTIVATION Sun to part shade and moist to seasonally moist, well-drained soil. Water to establish. Prefers a cool, moist environment. Shade tolerant, especially when young. A lovely large evergreen tree that can be planted as a specimen, windbreak, or screen. Give it ample space. Slow growing. Mulch.

Engelmann spruce does not favor maritime climates and ranges farther east than Sitka spruce, making it a better option for inland gardens. However, since it is naturally found in mountainous areas and at higher elevations, it prefers cool, moist soils and does not do well in hot, dry locations without supplemental water. This is a long-lived evergreen with a dense, slender habit and conical crown that can grow to well over 100 ft. tall but tends to remain shorter in cultivation. It has stiff, blue-green needles typical of spruces. Fresh new growth can be used to flavor beer. Many selections of Engelmann spruce are commonly available including dwarf forms.

Engelmann spruce is an attractive evergreen with stiff, blue-green needles that grows in mountainous areas.

Picea sitchensis • Pinaceae

Sitka spruce

HABITAT/RANGE Forests and coastal areas. Alaska to California mainly at low elevations. Grows west of the Cascade Mountains.

SEASONAL INTEREST Evergreen.

WILDLIFE VALUE Wind pollinated. Attracts beneficial insects. Butterfly and moth host plant. Seeds eaten by birds and small mammals. Provides critical habitat, cover, and nesting sites. Shades waterways and helps create healthy fish habitat. Supports epiphytic life. Needles eaten by grouse. Deer may browse new growth.

CULTIVATION Sun to part shade and wet to moist or seasonally moist, humus-rich, well-drained soil. Can grow in sandy soils and tolerates salt spray. Water to establish. Tolerates seasonally dry conditions once established if properly sited. Prefers cool, moist locations. A good specimen, windbreak, or screen for gardens west of the Cascades. Mulch.

Sitka spruce is an evergreen tree that can grow to great heights and girth in the moist, cool environs of maritime forests. It supports a large amount of life in its canopy, from small epiphytic plants to large birds like bald eagles. Beneath its towering crown, Sitka spruce provides critical habitat for all the creatures of the forest, rivers, and coastline. It is a handsome evergreen with stiff, green to bluish green foliage. In spring branch tips are adorned with tufts of soft new growth, which is high in vitamin C and can be used to flavor beer, syrups, and even ice cream. One of our largest conifers, it has the potential to grow over 200 ft. tall and 20 to 40 ft. wide in ideal conditions.

Sitka spruce has stiff needles that are silvery blue on the undersides and green on top.

Pinus contorta var. *contorta* • Pinaceae

Shore pine

HABITAT/RANGE Coastal areas, bogs, and mountain foothills. Alaska to Northern California at low elevations. Grows west of the Cascade Mountains.

SEASONAL INTEREST Orange-red pollen cones in spring to early summer. Long-lasting seed cones. Evergreen.

WILDLIFE VALUE Wind pollinated. Attracts beneficial insects. Butterfly larval host. Seeds eaten by birds and small mammals. Birds glean insects from trees. Provides cover and nesting sites. Stabilizes shoreline habitats. Browsed by porcupine. Deer resistant.

CULTIVATION Full sun and wet to seasonally dry soil. Shade intolerant. Adaptable to a variety of soil types including infertile and sandy soils. Tolerates salt spray. Water to establish. Drought tolerant once established but benefits from occasional supplemental water. Good for small spaces and containers. A lovely specimen for rock gardens and formal landscapes. Useful for screening and shoreline stabilization. Mulch.

Shore pine is a coastal species that has tenaciously eked out a niche in windswept environments, allowing nature's forces to twist and shape it into dwarfed and contorted forms. It has short needles in sets of two and grows 10 to 50 ft. tall and 10 to 40 ft. wide depending on site conditions. It produces elongated buds of new growth called candles in spring. "Candling" is the process of pinching off new growth, which is the best way to shape this beautifully structured tree. Another variation of the species, lodgepole pine (*Pinus contorta* var. *latifolia*), grows exceedingly straight and tall in mountainous inland areas.

Shore pine often has a dwarfed and contorted form.

Pinus monticola • Pinaceae

Western white pine

HABITAT/RANGE Mixed-conifer forests, mountainous areas, moist valleys, and open slopes. British Columbia to California and east to northern Idaho and Montana. Low to high elevations. Grows on both sides of the Cascade Mountains.

SEASONAL INTEREST Long, hanging cones. Evergreen.

WILDLIFE VALUE Wind pollinated. Attracts beneficial insects. Butterfly larval host. Seeds eaten by birds and small mammals. Provides cover, nesting sites, and nesting materials. Occasionally browsed by deer and elk.

CULTIVATION Sun to light shade and moist to seasonally dry soil. Grows in a variety of soil types. Tolerates poor soils. Water to establish. Drought tolerant once established if properly sited. Do not plant near currants or gooseberries (*Ribes* spp.), which are the alternate host for white pine blister rust. Rust resistant strains of this tree are available. May drip pitch in hot weather, place away from walkways. Mulch.

Western white pine is a handsome evergreen with bluish green needles in bundles of five. Able to grow over 100 ft. tall, its bark is silvery gray when young, becoming reddish and distinctly checkered with age. It tolerates a variety of soil conditions, an adaptation allowing it to grow in areas with less competition from other trees. The only thing it hasn't adapted well to is a devastating blow to its population from overharvesting and white pine blister rust, a fungus of Eurasian origin that infects five-needled pines. If you have the space, consider giving this beautiful conifer refuge in your landscape.

Western white pine has a bluish hue to its foliage and silvery gray bark when young.

Pinus ponderosa • Pinaceae

Ponderosa pine

HABITAT/RANGE Dry, open areas, open forests, and mountain slopes. British Columbia to Mexico and east to the Dakotas. Low to high elevations. Grows chiefly east of the Cascade Mountains with populations also west of the mountains.

SEASONAL INTEREST Large cones. Evergreen.

WILDLIFE VALUE Important for wildlife. Wind pollinated. Attracts beneficial insects. Butterfly larval host. Birds glean insects from trees. Seeds eaten by birds and mammals. Provides forage, cover, nesting sites, and nesting materials. Occasionally browsed by deer and elk.

CULTIVATION Sun to part shade and well-drained soil. Grows in a variety of soil types. Tolerates poor soils. Water to establish. Drought tolerant once established. Prefers dry summers. Best placed away from buildings as it sheds cones with sharp barbs and needles, which may create a fire hazard. Mulch.

Ponderosa pine is a true pillar of the community in vast parts of the inland West. Capable of creating productive habitat in harsh environments and towering over most of the plants it shares a home with, this large, stately evergreen can eventually grow to over 200 ft. tall and more than 25 ft. wide. It has long needles, usually in bunches of three, and large, prickly cones. One of the most stunning aspects of this handsome tree is its bark, which flaunts tonal, flaking puzzle-piece configurations in shades of cinnamon and burnt umber. This species is comprised of somewhat geographically divided, and taxonomically debated, variations adapted to different climates. For that reason, it is important to use locally sourced plant material.

Ponderosa pine is a stately, drought-tolerant evergreen with colorful puzzle-piece bark.

Populus tremuloides • Salicaceae

Quaking aspen

HABITAT/RANGE Meadows, prairies, forest openings, mountain slopes, riparian areas, and disturbed sites. The most widely distributed tree in North America. Alaska to Mexico and east to the Atlantic Coast. Low to high elevations. Grows on both sides of the Cascade Mountains.

SEASONAL INTEREST Gorgeous fall color. Beautiful bark. Deciduous.

WILDLIFE VALUE Larval host to many beautiful butterflies. Attracts beneficial insects. Birds glean insects from trees. Sapsuckers mine sap, leaving holes other species feed from; this is an important food source for migrating hummingbirds in spring. Wind pollinated. Honey bees collect resin to make propolis. Provides cover and nesting sites. Browsed by deer, elk, beaver, and other animals.

CULTIVATION Sun to light shade and moist to seasonally moist soil. Grows in a variety of soil types. Water to establish. Tolerant of seasonally dry conditions once established but prefers moist sites. Spreads vigorously; not appropriate for small spaces. Roots can damage pipes and foundations. Mulch.

Quaking aspen is an incredibly attractive deciduous tree with smooth white bark that puts on picturesque displays of fall foliage with its spade-shaped leaves that "tremble" in the wind. Trees grow quickly, reaching 20 to 60 ft. tall or more. A clonal species, it spreads to form extensive colonies, making it one of the largest, and oldest, organisms on Earth. While individual trees may be relatively short lived, colonies are practically immortal, cloning themselves indefinitely in stable sites. Aspen forests create a mecca for birds and its benefits to wildlife are innumerable; a true blend of beauty and biodiversity.

Quaking aspen is known for its beautiful fall foliage and smooth white bark.

Populus trichocarpa • Salicaceae

Black cottonwood

HABITAT/RANGE Riparian areas, floodplains, meadow edges, moist forests, canyon bottoms, and disturbed areas. Alaska to California and east to Montana at low to high elevations. Grows on both sides of the Cascade Mountains.

SEASONAL INTEREST Cottony seeds early summer. Fall color. Deciduous.

WILDLIFE VALUE Larval host to many beautiful butterflies. Attracts beneficial insects. Birds glean insects from trees. Sapsuckers mine sap, leaving holes other species feed from; this is an important food source for hummingbirds. Wind pollinated. Honey bees collect resin to make propolis. Provides cover and nesting sites. Helps create healthy riparian habitat. Browsed by deer, elk, beaver, and other animals.

CULTIVATION Sun to light shade and moist soil. Grows in a variety of soil types. Water to establish. Tolerates seasonally dry conditions once established if properly sited. Tolerates flooding. Roots can damage pipes and foundations. Excellent for wildlife and riparian areas. Cuttings root readily; can be established in riparian areas using live stakes. Useful for windbreaks. Mulch.

Black cottonwood is not the tree for small spaces or formal plantings but its benefits to wildlife are abundant. If you have a moist spot along a meadow or stream you want to shade and stabilize, this is your tree. While supporting birds, bees, and butterflies you can enjoy the golden hues of the fall foliage, as well as the sweet smell of its resin in spring as new leaves emerge. This fast-growing deciduous tree grows as much as 150 ft. tall and 30 ft. wide. Previously classified as *Populus balsamifera* subsp. *trichocarpa*.

Black cottonwood has lovely fall color.

Prunus emarginata • Rosaceae

Bitter cherry

HABITAT/RANGE Open woods, rocky slopes, streambanks, thickets, and disturbed areas. British Columbia to California and east to the Rocky Mountains at low to high elevations. Grows on both sides of the Cascade Mountains.

SEASONAL INTEREST White flowers in spring to early summer relative to elevation. Red fruits. Fall color. Deciduous.

WILDLIFE VALUE Attracts bees, bumble bees, butterflies, hummingbirds, and other pollinators. Butterfly and moth larval host. Attracts beneficial insects. Birds glean insects from plants. Fruits eaten by birds and mammals. Provides cover and nesting sites. Browsed by deer and elk.

CULTIVATION Sun to part shade and moist to seasonally dry, well-drained soil. Water to establish. Drought tolerant once established if properly sited. Spreads by suckers. Plant along woodland edges and in hedgerows. Great for pollinator and bird habitat. Mulch.

Prunus emarginata has an ethereal quality in spring when it covers itself in clusters of white flowers that look fuzzy due to their long stamens. Pollinators of all shapes and sizes rely on these flowers as an early-season source of pollen and nectar. The pollinator frenzy ensures the production of small, bright to dark red stone fruits, which are too bitter to be eaten and best left to the birds who will gladly gobble them up. With an upright habit, it grows anywhere from 6 to 50 ft. tall and 5 to 30 ft. wide depending on site conditions. This small deciduous tree with shrublike tendencies can spread to form thickets, making it best suited for edges and hedges.

Bitter cherry produces clouds of white flowers that feed a variety of pollinators and fruits that feed birds and other wildlife.

Prunus virginiana • Rosaceae

Choke cherry

HABITAT/RANGE Open forests, forest edges, thickets, riparian areas, coastal areas, rocky slopes, and roadsides. British Columbia to California and east to the Atlantic Coast at low to high elevations. Grows on both sides of the Cascade Mountains.

SEASONAL INTEREST Creamy white flowers in late spring to mid-summer depending on location. Red to purple fruits. Fall color. Deciduous.

WILDLIFE VALUE Attracts bees, butterflies, hummingbirds, and other pollinators. Larval host for many species of butterflies and moths. Attracts beneficial insects. Birds glean insects from plants. Fruits eaten by birds and mammals. Provides cover and nesting sites. Browsed by deer and elk.

CULTIVATION Sun to part shade and moist to seasonally dry, well-drained soil. Water to establish. Drought tolerant once established if properly sited. Heat tolerant. Prefers some moisture and afternoon shade in arid areas. Can spread by suckers. Grows well in containers. Great for just about any sunny garden or habitat planting. Mulch.

Gorgeous in bloom, *Prunus virginiana* is a handsome ornamental that benefits wildlife. Its profusions of elongated clusters of creamy white flowers make it popular with both gardeners and pollinators. Many beautiful butterflies, including swallowtails and admirals, feed on the plant and its flowers. Choke cherry's red to purple stone fruits are edible but bitter and potentially toxic; remove the seeds and cook or dry the fruit before eating. If you don't eat them, the birds happily will. This small deciduous tree to large shrub grows 10 to 30 ft. tall and 10 to 20 ft. wide.

Choke cherry has long, showy clusters of creamy white flowers.

Pseudotsuga menziesii • Pinaceae

Douglas fir

HABITAT/RANGE Moist to dry forests and open areas from British Columbia to California and east to the Rocky Mountains. Low to high elevations. Grows on both sides of the Cascade Mountains.

SEASONAL INTEREST Evergreen.

WILDLIFE VALUE Important tree for wildlife. Wind pollinated. Butterfly and moth host plant. Birds glean insects from trees. Seeds eaten by birds and small mammals. Provides cover and nesting sites. Browsed by grouse. Browsed by deer and elk only when more palatable forage is scarce.

CULTIVATION Sun to part shade and moist to seasonally dry, humus-rich, well-drained soil. Does well in average or slightly acidic soils. Water to establish. Drought tolerant once established if properly sited. Benefits from some supplemental water in arid areas. Large; give it plenty of space and head room. Mulch.

Douglas fir is a keystone species in the evergreen expanses of the Pacific Northwest. It is elemental in the lives of the region's inhabitants and shelters a wide diversity of life. This large, relatively fast-growing tree typically reaches around 100 ft. high and 30 ft. wide, though it can grow much taller. Plant it if you need some shade and want to foster a forest. Two variations are found in the region. Coast Douglas fir (*Pseudotsuga menziesii* var. *menziesii*) makes up the population west of the Cascades but also grows east of the mountains. Rocky Mountain Douglas fir (*P. menziesii* var. *glauca*) grows east of the Cascades through the Rocky Mountains and has green to blue-green foliage.

Douglas fir is a fundamental component of the Pacific Northwest's evergreen forests.

Quercus garryana var. garryana • Fagaceae

Oregon white oak

HABITAT/RANGE Oak woodlands and savannas, mixed forests, prairies, meadows, riparian areas, rocky slopes, and bluffs. British Columbia to California at low to mid elevations. Grows mainly from the east base of the Cascade Mountains to the coast.

SEASONAL INTEREST Flowers in spring. Acorns late summer to fall. Fall color. Deciduous.

WILDLIFE VALUE Essential to many species. Supports and hosts numerous beneficial insects, butterflies, and moths. Wind pollinated. Birds, deer, elk, bear, squirrels, and a wide diversity of wildlife rely on these trees for food and shelter.

CULTIVATION Sun to light shade and moist to seasonally dry, well-drained soil. Grows in a variety of soil types. Water to establish. Drought tolerant once established if properly sited. Slow growing. Protect young plants from deer. Mulch.

Oregon white oaks ARE habitat. They support a staggering amount of insects and wildlife, and provide food and shelter for some of the region's most sensitive species. Their leaves are late to emerge, and many wildflowers are adapted to the availability of spring sunshine yet protection from the harsh summer sun that oaks provide. Oregon white oak woodlands are in peril, with around 90 percent cleared and converted to other uses. Homeowners and neighborhoods can help by planting and preserving oak trees. It will take some time, but they can eventually grow 25 to 100 ft. tall and 20 to 60 ft. wide. Plant this tree for future generations. Other variations of this species are more shrublike and the genus is particularly diverse in California.

Oregon white oaks are stately trees that form the backbone of some of the most important and imperiled habitats in the region.

Salix spp. • Salicaceae

Willow

HABITAT/RANGE Moist sites, riparian areas, wetlands, meadows, forests, subalpine slopes, and disturbed areas. Many species are found throughout the region.

SEASONAL INTEREST Blooms early spring. Fall color. Deciduous. Brightly colored twigs of some species are attractive in winter.

WILDLIFE VALUE Early-season pollen and nectar source for bees, butterflies, and other pollinators. Attracts beneficial insects. Important butterfly larval host. Specialist bee host. Enhances riparian habitat. Provides cover and forage. Browsed by deer.

CULTIVATION Sun to part shade and wet to seasonally moist soil. Water to establish and continue to provide supplemental water as needed. Some species tolerate seasonally dry conditions if properly sited. May spread by rhizomes. Plant in riparian areas, bioswales, rain gardens, woodland edges, and pollinator habitat. Use male plants if planting for bees as they produce pollen. Cuttings root easily; shoreline stabilization is quickly achieved using bundles of live stakes (willow wattles) placed horizontally and mostly buried in wet soil. Mulch.

Willows stabilize riparian areas, provide an early-season food source for bees, and are a larval host for some of our most beautiful butterflies. Both male and female plants produce flowering spikes called catkins, which can be soft and furry before blooming. Many species are native to the region, some develop into small trees while others grow as shrubs. Popular species for landscaping include Scouler's willow (*Salix scouleriana*), a drought-tolerant species found in forests and mesic habitat; Pacific willow (*S. lasiandra*), which tolerates flooding; and coyote willow (*S. exigua*), which grows near waterways east of the Cascades.

Willows love wet places.

Sambucus cerulea • Adoxaceae/Viburnaceae

Blue elderberry

HABITAT/RANGE Forest edges, fields, streambanks, rocky slopes, and moist areas in dry habitat. Southern British Columbia south through California and east to the Rocky Mountains at low to high elevations. Grows on both sides of the Cascade Mountains.

SEASONAL INTEREST White flowers late spring to midsummer. Edible blue fruits. Deciduous.

WILDLIFE VALUE Attracts bees, butterflies, hummingbirds, and other pollinators. Butterfly and moth larval host. Attracts beneficial insects. Fruits eaten by birds and wildlife. Provides cover and nesting sites. Pithy stems used by cavity-nesting bees like mason bees. Occasionally browsed by deer and elk.

CULTIVATION Sun to light shade and moist to seasonally dry, humus-rich, well-drained soil. Water to establish. Tolerates seasonally dry conditions once established but benefits from deep, infrequent summer water. May die back to the ground in winter its first few years. Grows quickly once established. When pruning leave at least 4 to 6 in. above the leaf node to create bee nesting sites. Worthy addition to farms, food forests, and backyard bird habitat. Mulch.

Blue elderberry is a beautiful deciduous tree or shrub that grows 8 to 30 ft. tall, usually less wide, with pinnately compound leaves, pithy branches that provide habitat for cavity-nesting bees, and fragrant, creamy white flowers in large flat-topped clusters that attract pollinators. Loved by birds, the berries are coated with a powdery yeast that makes them look blue. They are edible to humans if cooked; however, most parts of the plant are toxic. There are many exotic elderberry cultivars commonly available; be sure to plant true natives. Classification of this species may vary.

Blue elderberry has attractive leaves, flowers, and fruit.

Sorbus scopulina • Rosaceae

Cascade mountain ash

HABITAT/RANGE Forests and forest edges, thickets, meadows, riparian areas, and rocky slopes in mountainous areas. Alaska to California and east through the Rocky Mountains at mid to high elevations. Grows in and on both sides of the Cascade Mountains.

SEASONAL INTEREST White flowers in late spring to summer relative to elevation. Red-orange fruits late summer to fall. Fall color. Deciduous.

WILDLIFE VALUE Attracts bees, butterflies, and other pollinators. Fruits eaten by birds and mammals. Provides cover. Browsed by deer and elk.

CULTIVATION Sun to part shade and moist to seasonally dry, humus-rich, well-drained soil. Water to establish. Drought tolerant once established if properly sited. Naturally grows at higher elevations where temperatures are cooler; benefits from afternoon shade and supplemental water in arid areas. Hot and dry conditions may stress plants making them vulnerable to disease. An attractive plant for woodland edges, edible landscapes, and hedgerows. Mulch.

Pinnately compound leaves give Cascade mountain ash a tropical look. These leaves are dark green and shiny on the upper surface, turning lovely shades of red, peach, or yellow in fall. Its large clusters of white flowers are showy and become even showier clusters of red-orange fruits. The fruits are edible to wildlife as well as humans and can be used in pies and preserves. Taking the form of either a small tree or shrub, it grows 5 to 15 ft. tall and 5 to 10 ft. wide. Sitka mountain ash (*Sorbus sitchensis*) is similar in appearance and preferred growing conditions.

Cascade mountain ash has showy flowers and red-orange fruits.

Taxus brevifolia • Taxaceae

Pacific yew

HABITAT/RANGE Moist forests from Alaska to California and east to Montana. Low to mid elevations. Grows on both sides of the Cascade Mountains.

SEASONAL INTEREST Fleshy, red fruits late summer to fall. Evergreen.

WILDLIFE VALUE Fruits eaten by birds. Provides evergreen cover and nesting sites. Browsed by deer, elk, and moose.

CULTIVATION Full to part shade and moist to seasonally moist, humus-rich, well-drained soil. Water to establish. Drought tolerant once established if properly sited. Slow growing. Plant in shady forest gardens. Mulch.

Pacific yew is a small evergreen tree that grows 20 to 50 ft. tall and 10 to 20 ft. wide, making it more of an understory plant than a canopy contributor. Its needles are arranged in two flat rows along the stems and have a somewhat airy look. The bark is scaly and peels off revealing a reddish to purplish inner bark. Plants are either male or female, with female trees producing seeds held in a fleshy, red, berrylike casing called an aril. While this tree is the celebrated source of an anticancer drug, it is highly toxic. The red flesh of the fruits is reportedly edible, but seeds within them are poisonous so leave them for the birds. This unique tree favors old growth forests, but as those are in ever shorter supply, a place in your landscaping will be a welcome refuge.

Pacific yew is a special tree with airy foliage and peeling bark.

Thuja plicata • Cupressaceae

Western red cedar

HABITAT/RANGE Moist sites, mixed forests, wetlands, riparian areas, and coastal areas. Alaska to Northern California and east to Montana at low to moderately high elevations. Grows on both sides of the Cascade Mountains.

SEASONAL INTEREST Evergreen.

WILDLIFE VALUE Wind pollinated. Seeds eaten by birds. Butterfly and moth larval host. Stabilizes soils, shades waterways, and helps create healthy habitat for aquatic life. Provides cover, nesting sites, and nesting materials. Browsed by deer and elk.

CULTIVATION Sun to part shade and moist to seasonally moist, humus-rich soil. Prefers slightly acidic soils and moist sites. Water to establish and continue to provide supplemental water as needed. Tolerant of seasonally dry conditions once established if properly sited. Needs afternoon shade and supplemental water in hot, dry areas. A beautiful tree for formal plantings, as well as riparian restoration and wildlife habitat. Mulch.

Western red cedar is a legendary evergreen utilized by people for medicine, floral decorations, rot resistant lumber, and even clothing. It has a lovely fragrance and pleasant structure with flat sprays of foliage, swooping branches, reddish bark, and buttressed trunks. A relative of arborvitae, it is useful for screening, as well as hedges when pruned correctly. The most sublime sight, however, is a fully grown cedar shading a streambank with ample space to spread out its branches. Though often growing 75 to 150 ft. tall and 25 to 50 ft. wide, it can grow much taller.

Western red cedar is a beautifully structured, moisture-loving evergreen.

Tsuga heterophylla • Pinaceae

Western hemlock

HABITAT/RANGE Moist forests from Alaska to Northern California. Grows mainly in and west of the Cascade Mountains, also from southeast British Columbia to northeast Washington, northern Idaho, and western Montana. Low to mid elevations.

SEASONAL INTEREST Evergreen.

WILDLIFE VALUE Wind pollinated. Attracts beneficial insects. Butterfly and moth larval host. Birds glean insects from trees. Seeds eaten by birds and mammals. Provides cover and nesting sites. Browsed by deer, elk, and other wildlife.

CULTIVATION Sun to shade and moist to seasonally dry, humus-rich, well-drained soil. Very shade tolerant but grows faster in sun. Water to establish. Tolerates seasonally dry conditions once established if properly sited. Grows best in cool, moist climates. Can be grown as a hedge with careful pruning. Relatively shallow roots make it susceptible to windthrow; place away from buildings in high wind areas. Mulch.

The narrow, drooping crown and lacy foliage of western hemlock is a familiar sight in forests west of the Cascades. Young seedlings of this evergreen tree readily colonize nurse logs, growing slowly in the moist understory shade. In spring and early summer its delicate-looking foliage becomes even more attractive when light green skirts of new needles appear along the tips of the branches. Growing 40 to 150 ft. tall and 20 to 30 ft. wide, mature trees offer cover and nesting sites to birds and wildlife including cavity-nesting species like woodpeckers and owls. Make sure to plant this handsome tree in a seasonally moist site with plenty of organic matter in the soil.

Western hemlock is a graceful evergreen with finely textured foliage.

Tsuga mertensiana • Pinaceae

Mountain hemlock

HABITAT/RANGE Forests, moist sites, and alpine slopes in mountainous areas. Alaska to California and east to Montana. Mid to high elevations, grows at sea level and in bogs in the northern portion of its range. Grows in and on both sides of the Cascade Mountains.

SEASONAL INTEREST Purple seed cones. Evergreen.

WILDLIFE VALUE Wind pollinated. Attracts beneficial insects. Butterfly and moth host plant. Birds glean insects. Seeds eaten by birds and small mammals. Provides cover and nesting sites. Browsed by grouse.

CULTIVATION Sun to part shade and wet to moist or seasonally moist, humus-rich, well-drained soil. Prefers some shade in hot climates. Grows best in moist places with cool summers. Water to establish and continue to supply supplemental water as needed. Tolerates seasonally dry conditions once established if properly sited. Slow and compact growth make it suitable for small spaces. Can be grown in containers. Used for bonsai. Susceptible to windthrow. Nice specimen for formal plantings, alpine gardens, and rock gardens. Mulch.

Mountain hemlock takes refuge in cool mountain areas where spartan conditions have molded it into a compact, slow-growing tree capable of bending to the will of heavy snowfall. Its short needles are often blue-green or silvery hued, and the seed cones are nearly purple when developing. The height and form of this tree are relative to site conditions. In alpine environments it is dwarfed and even shrublike, but it can grow over 100 ft. tall, though usually no more than 30 ft. tall in cultivation. This is a popular and much-loved ornamental.

Mountain hemlock is a rugged evergreen capable of growing in austere mountain conditions.

Plants for Specific Purposes

Plants for Shade

- *Acer circinatum*
- *Achlys* spp.
- *Actaea rubra*
- *Anemone deltoidea*
- *Aquilegia formosa*
- *Aruncus dioicus*
- *Asarum caudatum*
- *Berberis nervosa*
- *Carex geyeri*
- *Cornus* spp.
- *Corydalis scouleri*
- *Delphinium trolliifolium*
- *Dicentra formosa*
- Ferns
- *Gaultheria shallon*
- *Heracleum maximum*
- *Heuchera micrantha*
- *Hydrophyllum tenuipes*
- *Linnaea borealis*
- *Lysichiton americanus*
- *Maianthemum* spp.
- *Nothochelone nemorosa*
- *Oemleria cerasiformis*
- *Oxalis oregana*
- *Petasites frigidus*
- *Polemonium* spp.
- *Prosartes hookeri*
- *Rhododendron macrophyllum*
- *Rubus spectabilis*
- *Sambucus racemosa*
- *Synthyris reniformis/ Veronica regina-nivalis*
- *Taxus brevifolia*
- *Tellima grandiflora*
- *Thalictrum occidentale*
- *Tolmiea menziesii*
- *Trillium ovatum*
- *Tsuga heterophylla*
- *Vaccinium* spp.
- *Vancouveria hexandra*
- *Viola glabella*

Drought-tolerant Plants

Many of the region's plants are adapted to dry summer conditions but these plants are particularly drought tolerant.

- *Achillea millefolium*
- *Achnatherum hymenoides/ Eriocoma hymenoides*
- *Allium acuminatum*
- *Amelanchier alnifolia*
- *Arbutus menziesii*
- *Arctostaphylos* spp.
- *Artemisia* spp.
- *Asclepias fascicularis*
- *Balsamorhiza* spp.
- *Berberis aquifolium*
- *Berberis repens*
- *Calocedrus decurrens*
- *Carex geyeri*
- *Ceanothus integerrimus*
- *Celtis reticulata*
- *Clarkia pulchella*
- *Delphinium nuttallianum*
- *Dichelostemma congestum*
- *Drymocallis glandulosa*
- *Elymus elymoides*
- *Ericameria nauseosa*
- *Erigeron linearis*
- *Eriogonum* spp.
- *Eriophyllum lanatum*
- *Eschscholzia californica*
- *Festuca idahoensis*
- *Gaillardia aristata*
- *Garrya fremontii*
- *Gilia capitata*
- *Helianthella uniflora*
- *Heterotheca villosa*
- *Holodiscus discolor*
- *Hydrophyllum capitatum*
- *Koeleria macrantha*
- *Lewisia rediviva*
- *Leymus cinereus*
- *Lomatium* spp.
- *Lupinus latifolius*

Drought-tolerant Plants (*continued*)

- *Lupinus leucophyllus*
- *Monardella odoratissima*
- *Opuntia* spp.
- *Penstemon* spp.
- *Phacelia hastata*
- *Philadelphus lewisii*
- *Phlox* spp.
- *Pinus ponderosa*
- *Prunus* spp.
- *Purshia tridentata*
- *Quercus garryana*
- *Ranunculus occidentalis*
- *Rhus* spp.
- *Ribes aureum* var. *aureum*
- *Rosa woodsii*
- *Sedum* spp.
- *Spiraea lucida*
- *Symphoricarpos* spp.
- *Triteleia grandiflora*

Plants for Wetlands and Riparian Areas

- *Adiantum aleuticum*
- *Alnus rubra*
- *Aruncus dioicus*
- *Carex obnupta*
- *Coreopsis tinctoria*
- *Cornus sericea*
- *Corydalis scouleri*
- *Darmera peltata*
- *Erythranthe guttata*
- *Fraxinus latifolia*
- *Helenium autumnale*
- *Heracleum maximum*
- *Juncus effusus* subsp. *pacificus*
- *Lupinus polyphyllus*
- *Lysichiton americanus*
- *Petasites frigidus*
- *Physocarpus capitatus*
- *Populus* spp.
- *Rubus spectabilis*
- *Rudbeckia occidentalis*
- *Sagittaria latifolia*
- *Salix* spp.
- *Schoenoplectus tabernaemontani*
- *Spiraea douglasii*
- *Stachys cooleyae*
- *Thuja plicata*
- *Typha latifolia*
- *Viola glabella*

Evergreen Plants

- *Abies* spp.
- *Arbutus menziesii*
- *Arctostaphylos* spp.
- *Armeria maritima*
- *Artemisia tridentata*
- *Asarum caudatum*
- *Berberis* spp.
- *Calocedrus decurrens*
- *Carex geyeri*
- *Ceanothus thyrsiflorus*
- *Garrya* spp.
- *Gaultheria shallon*
- *Juncus effusus* subsp. *pacificus*
- *Morella californica/ Myrica californica*
- *Opuntia* spp.
- *Paxistima myrsinites*
- *Penstemon davidsonii*
- *Penstemon fruticosus*
- *Picea* spp.
- *Pinus* spp.
- *Polystichum munitum*
- *Pseudotsuga menziesii*
- *Rhododendron macrophyllum*
- *Sedum* spp.
- *Struthiopteris spicant/ Blechnum spicant*
- *Taxus brevifolia*
- *Thuja plicata*
- *Tsuga* spp.
- *Vaccinium ovatum*

Low-Growing Groundcovers

- *Achlys* spp.
- *Arctostaphylos uva-ursi*
- *Asarum caudatum*
- *Berberis repens*
- *Campanula rotundifolia*
- *Carex geyeri*
- *Clinopodium douglasii*
- *Cornus unalaschkensis*
- *Epilobium canum*
- *Fragaria* spp.
- *Geum triflorum*
- *Linnaea borealis*
- *Maianthemum dilatatum*
- *Oxalis oregana*
- *Phlox diffusa*
- *Prunella vulgaris* var. *lanceolata*
- *Symphoricarpos mollis*
- *Tolmiea menziesii*
- *Vancouveria hexandra*
- *Viola glabella*

Aromatic Plants

- *Abies grandis*
- *Agastache urticifolia*
- *Allium* spp.
- *Artemisia* spp.
- *Asarum caudatum*
- *Asclepias speciosa*
- *Berberis aquifolium*
- *Calocedrus decurrens*
- *Ceanothus integerrimus*
- *Clinopodium douglasii*
- *Lomatium papilioniferum*
- *Lysichiton americanus*
- *Monardella odoratissima*
- *Morella californica/ Myrica californica*
- *Philadelphus lewisii*
- *Purshia tridentata*
- *Ribes aureum* var. *aureum*
- *Rosa* spp.
- *Stachys cooleyae*
- *Thuja plicata*
- *Valeriana sitchensis*

Deer-Resistant Plants

Plants in their natural habitat may be resistant to deer, but if planted where they don't normally occur they may become a deer delicacy. When staple food sources are lacking, deer are forced to eat plants they don't regularly enjoy. Deer and elk may rely on certain plants for winter forage but ignore them in other seasons. For these reasons it is difficult to provide a definitive list of deer-resistant plants. General deer resistance or desirability is noted when possible in the plant profiles but there are no guarantees.

Great Hummingbird Plants

- *Agastache urticifolia*
- *Aquilegia formosa*
- *Arbutus menziesii*
- *Arctostaphylos* spp.
- *Berberis* spp.
- *Castilleja* spp.
- *Chamaenerion angustifolium*
- *Delphinium* spp.
- *Dicentra formosa*
- *Dichelostemma congestum*
- *Epilobium canum*
- *Erythranthe* spp.
- *Ipomopsis aggregata*
- *Lilium* spp.
- *Lonicera* spp.
- *Lupinus* spp.
- *Nothochelone nemorosa*
- *Oemleria cerasiformis*
- *Penstemon* spp.
- *Populus* spp.
- *Ribes* spp.
- *Rubus spectabilis*
- *Stachys cooleyae*
- *Symphoricarpos* spp.
- *Typha latifolia*
- *Vaccinium* spp.

Great Butterfly Plants

- *Achillea millefolium*
- *Agastache urticifolia*
- *Alnus rubra*
- *Anaphalis margaritacea*
- *Angelica lucida*
- *Arctostaphylos* spp.
- *Asclepias* spp.
- *Balsamorhiza* spp.
- *Berberis* spp.
- *Camassia* spp.
- *Ceanothus* spp.
- *Crataegus douglasii*
- *Deschampsia cespitosa*
- *Dichelostemma congestum*
- *Ericameria nauseosa*
- *Erigeron* spp.
- *Eriogonum* spp.
- *Eriophyllum lanatum*
- *Festuca idahoensis*
- *Gilia capitata*
- *Helenium autumnale*
- *Helianthus nuttallii*
- *Heracleum maximum*
- *Heterotheca villosa*
- *Holodiscus discolor*
- *Koeleria macrantha*
- *Lilium columbianum*
- *Lomatium* spp.
- *Lupinus* spp.
- *Monardella odoratissima*
- *Petasites frigidus*
- *Philadelphus lewisii*
- *Phlox* spp.
- *Populus* spp.
- *Prunus* spp.
- *Quercus garryana*
- *Ribes* spp.
- *Salix* spp.
- *Sambucus* spp.
- *Sedum* spp.
- *Sidalcea* spp.
- *Solidago* spp.
- *Spiraea* spp.
- *Symphyotrichum foliaceum*
- *Valeriana sitchensis*

Plants that Provide Fruits for Birds

- *Amelanchier alnifolia*
- *Arbutus menziesii*
- *Berberis* spp.
- *Celtis reticulata*
- *Cornus* spp.
- *Crataegus douglasii*
- *Fragaria* spp.
- *Frangula purshiana*
- *Gaultheria shallon*
- *Lonicera* spp.
- *Maianthemum* spp.
- *Oemleria cerasiformis*
- *Prunus* spp.
- *Rhus* spp.
- *Ribes* spp.
- *Rosa* spp.
- *Rubus* spp.
- *Sambucus* spp.
- *Sorbus* spp.
- *Symphoricarpos* spp.
- *Vaccinium* spp.
- *Viburnum* spp.

References and Resources

Included here are some of the references and resources we use to learn about the plants of the Pacific Northwest. We encourage you to use them as well. Botany is never boring, and our understanding of plants is always changing. Many of the botanical manuals that apply to the region have been recently revised or published for the first time, offering a lot of new information. For both the lifelong learners and those just meeting these plants, the following resources will help you better understand the region's flora and fauna.

Flora

Arno, Stephen F., and Ramona P. Hammerly. 2007. *Northwest Trees: Identifying and Understanding the Region's Native Trees*. Seattle, WA: Mountaineers Books.

DeBolt, Ann, Roger Rosentreter, and Valerie Geertson, eds. 2003. *Landscaping with Native Plants of the Intermountain Region*. Technical Reference 1730-3. Boise, ID: U.S. Department of the Interior, Bureau of Land Management.

Hitchcock, C. Leo, and Arthur Cronquist. 2018. *Flora of the Pacific Northwest: An Illustrated Manual*. 2nd ed. Edited by D. E. Giblin, B. S. Legler, P. F. Zika, and R. G. Olmstead. Seattle, WA: University of Washington Press.

Kruckeberg, Arthur R. 1996. *Gardening with Native Plants of the Pacific Northwest: Second Edition, Revised and Enlarged*. Seattle, WA: University of Washington Press.

Meyers, S. C., T. Jaster, K. E. Mitchell, T. Harvey, and L. K. Hardison, eds. 2015. *Flora of Oregon. Volume 1: Pteridophytes, Gymnosperms, and Monocots*. Botanical Research Institute of Texas, Fort Worth.

Meyers, S. C., T. Jaster, K. E. Mitchell, T. Harvey, and L. K. Hardison, eds. 2020. *Flora of Oregon. Volume 2: Dicots A–F*. Botanical Research Institute of Texas, Fort Worth.

Ogle, D., D. Tilley, J. Cane, L. St. John, K. Fullen, M. Stannard, and P. Pavek. 2017. Plants for Pollinators in the Intermountain West. *Plant Materials Technical Note* No. 2A. Revision. Boise, ID; Salt Lake City, UT; Spokane, WA: USDA Natural Resources Conservation Service.

Parish, Roberta, Ray Coupé, and Dennis Lloyd. 1996. *Plants of Southern Interior British Columbia and the Inland Northwest*. Vancouver, BC: Lone Pine.

Pendergrass, Kathy, Mace Vaughn, and Joe Williams. 2008. Plants for Pollinators in Oregon. *Plant Materials Technical Note* No. 13. Portland, OR: USDA Natural Resources Conservation Service.

Pojar, Jim, and Andy MacKinnon. 2004. *Plants of the Pacific Northwest Coast*. Rev. ed. Vancouver, BC: Lone Pine.

Pojar, Jim, and Andy MacKinnon. 2013. *Alpine Plants of the Northwest*. Edmonton, Alberta: Lone Pine.

Robson, Kathleen A., Alice Richter, and Marianne Filbert. 2008. *Encyclopedia of Northwest Native Plants for Gardens and Landscapes*. Portland, OR: Timber Press.

Roché, C. T., R. E. Brainerd, B. L. Wilson, N. Otting, R. C. Korfhage. 2019. *Field Guide to the Grasses of Oregon and Washington*. Corvallis, OR: University of Oregon Press.

Stark, Eileen M. 2014. *Real Gardens Grow Natives*. Seattle, WA: Skipstone.

Strickler, Dee. 1997. *Northwest Penstemons*. Columbia Falls, MT: Flower Press.

Turner, Mark, and Ellen Kuhlmann. 2014. *Trees and Shrubs of the Pacific Northwest*. Portland, OR: Timber Press.

Turner, Mark, and Phyllis Gustafson. 2006. *Wildflowers of the Pacific Northwest*. Portland, OR: Timber Press.

Fauna

LaPenta, Dante. 22 October 2018. Biodiversity for the Birds. *UDaily*. udel.edu/udaily/2018/october/non-native-plants-birds-insects-washington-chickadee-desiree-narango-doug-tallamy/.

Link, Russell. 1999. *Landscaping for Wildlife in the Pacific Northwest*. Seattle, WA: University of Washington Press.

Narango, Desirée L., Douglas W. Tallamy, and Peter P. Marra. 2018. Nonnative Plants Reduce Population Growth of an Insectivorous Bird. *Proceedings of the National Academy of Sciences* 115 (45): 11549–54. doi.org/10.1073/pnas.1809259115.

Pyle, Robert Michael, and Caitlin C. LaBar. 2018. *Butterflies of the Pacific Northwest*. Portland, OR: Timber Press.

Tallamy, Douglas W. 2009. *Bringing Nature Home: How You Can Sustain Wildlife with Native Plants*. Portland, OR: Timber Press.

Wheeler, Justin. 21 November 2017. Picking Plants for Pollinators: The Cultivar Conundrum. *Xerces Blog*. xerces.org/blog/cultivar-conundrum.

Xerces Society. 2011. *Attracting Native Pollinators: Protecting North America's Bees and Butterflies*. North Adams, MA: Storey Publishing.

Xerces Society. 2016. *Gardening for Butterflies*. Portland, OR: Timber Press.

Xerces Society. 2020. Nesting and Overwintering Habitat for Pollinators and Other Beneficial Insects. xerces.org/sites/default/files/publications/18-014.pdf.

Online Resources

A wealth of information awaits you online. Many organizations provide searchable databases with information about native plants, pollinators, and wildlife. Some provide nursery listings. Local nurseries may also post regionally appropriate information on growing native plants.

Audubon Society, Native Plants Database
audubon.org/native-plants

Biota of North America Program
bonap.org

Budburst, Nativars Research Project
budburst.org/nativars

BugGuide
bugguide.net

**Burke Herbarium Image Collection
(range and description of Washington flora)**
biology. burke.washington.edu/herbarium
/imagecollection.php

Butterflies and Moths of North America
butterfliesandmoths.org

Calflora
calflora.org

California Native Plant Society
cnps.org

**Calscape (plant database includes Lepidoptera
host plant information)**
calscape.org

Consortium of Pacific Northwest Herbaria
pnwherbaria.org

Cornell Lab of Ornithology, All About Birds
allaboutbirds.org

Electronic Atlas of the Flora of British Columbia
ibis.geog.ubc.ca/biodiversity/eflora

Flora and Fauna Northwest
science.halleyhosting.com

Flora of North America
floranorthamerica.org

Idaho Fish and Game, Idaho Species
idfg.idaho.gov/species

Idaho Native Plant Society
idahonativeplants.org

**King County, Native Plant Resources
for the Pacific Northwest**
kingcounty.gov/services/environment/stewardship
/nw-yard-and-garden/native-plant-resources-nw.aspx

Lady Bird Johnson Wildflower Center
wildflower.org/plants-main

Missouri Botanical Garden, Plant Finder
missouribotanicalgarden.org/plantfinder
/plantfindersearch.aspx

National Wildlife Federation, Native Plant Finder
nwf.org/NativePlantFinder

Native Plant Network
npn.rngr.net

Native Plant Society of British Columbia
npsbc.wordpress.com

Native Plant Society of Oregon
npsoregon.org

Native Plants PNW
nativeplantspnw.com

Native Seed Network
appliedeco.org/restoration/nativeseednetwork

OregonFlora (Oregon plants, includes range and gardening information)
oregonflora.org

Oregon State University, Landscape Plants
landscapeplants.oregonstate.edu

PlantNative (provides a nursery directory by state)
plantnative.org

Pollinator Partnership, Ecoregional Planting Guides
pollinator.org/guides

University and Jepson Herbaria
ucjeps.berkeley.edu

USDA Fire Effects Information System (FEIS)
feis-crs.org/feis

USDA Plant Materials Program, Publications and Resources
nrcs.usda.gov/wps/portal/nrcs/detail/plantmaterials
  /technical/publications

USDA Plants Database
plants.sc.egov.usda.gov

Washington Native Plant Society
wnps.org

Wildflower Search
wildflowersearch.org

Xerces Society
xerces.org

Conservation Districts

Local conservation districts provide excellent resources and support for planting native plants. Many also host annual native plant sales.

Oregon Soil and Water Conservation Districts
oregon.gov/oda/programs/naturalresources/swcd/pages
  /swcd.aspx

Washington Conservation Districts
scc.wa.gov/conservation-district-map

Idaho Association of Soil Conservation Districts
iascd.org

California Association of Resource Conservation Districts
carcd.org

Backyard Habitat Certification Programs

Backyard Habitat Certification Program (Portland Area)
backyardhabitats.org

National Wildlife Federation
nwf.org/backyard

Washington Deptartment of Fish and Wildlife, Habitat at Home
wdfw.wa.gov/species-habitats/living/backyard

Acknowledgments

We extend deep heartfelt thanks to our family and friends who have helped and encouraged us in the writing of this book and our work with native plants. Your love and support are essential in our lives, and we thank each of you. Our gratitude for the plants and animals who have been our teachers is also deep. Special appreciation goes to Sheila Ford Richmond for her efforts to create beautiful spaces for plants, wildlife, and people at Willow Ponds in Hood River, OR, where many of our photos were taken. A round of applause to Scott Hoelscher, Courtney Vengarick, and all the staff and volunteers at Leach Botanical Garden in Portland, OR, for the sanctuary they have created. It was a beautiful place to visit during the pandemic and we thank them for letting us share photographs taken there. We are also thankful to share pictures of the native plant demonstration garden at the Hegewald Center in Stevenson, WA, and commend the wonderful people at the Skamania County Noxious Weed Control Program and the volunteers who make that garden thrive. Our sincere gratitude to everyone at Columbia Land Trust for protecting and preserving critical habitat and sensitive species, as well as supporting and promoting backyard habitat programs. Thanks also to the Columbia Gorge Discovery Center & Museum in The Dalles, OR, as well as Mosier Company and Roots Hair Salon in our hometown Mosier, OR, for maintaining beautiful spaces filled with native plants. Super kudos to Megan Farrell and Bethany Womack for their editorial skills and feedback, which was helpful and vital. Appreciation also goes to Cathy Flick for her feedback and efforts in bird conservation. Thanks to Joy Markgraf, Matt Wenner, Ben and Michael Anderson-Nathe, and Scott and Sherry Pendarvis for the sanctuaries they steward where we found many happy plants to photograph. Thanks also to Giselle Kennedy Lord for the photography tips and tricks. Our botanical community is made up of really wonderful people with whom we enjoy doing conservation work and riddling out plant identification in the field. We are thankful for all the amazing botanists and native plant advocates we have had the opportunity to know and work with over our many years propagating, planting, and protecting native plants. In particular we would like to thank Krista Thie, Carolyn Wright, Paul Slichter, and Keith Karoly for their help and expertise along the way, we have learned much from each of you. A nod to the authors, gardeners, nurseries, and plant people actively sharing information about these plants and their experiences growing them. Lastly, we would like to acknowledge all the remarkable people we have had the pleasure to meet through our work and nursery who are out there planting native plants. It takes a village, and we thank you for being a part of growing a better future for all.

Photography Credits

Credit and thanks go to our niece, Hazel Womack, for providing us with her perspective on great purple monkeyflower in the introduction! All other photographs were taken by the authors, but credit is due to the flora and fauna of our home and nursery, the precious and spectacular wild places of the Pacific Northwest, and the gardens of individuals, communities, businesses, and botanical gardens mentioned in our acknowledgments.

From the pollinator gag reel.

Index

© Andrew Merritt

© Kristin Currin

Kristin Currin is the cofounder of Humble Roots Nursery, a native plant nursery in the Columbia River Gorge recognized for its efforts in sustainability and promoting native plants. Kristin noticed the beauty of native plants at an early age while exploring the fields and forests of her childhood in Ohio. Traveling and studying biology, agroforestry, and ethnobotany, she came to the Pacific Northwest and cultivated her passion for native plants while working with them in nurseries and in the field. She currently lives off-grid in Oregon with her husband, Andrew, where they run the nursery, ethically propagating many important species of native plants. Their labor of love has involved them with innumerable native plant endeavors including pollinator and conservation plantings of all shapes and sizes, school gardens, backyard habitats, restoration projects, and rare plant conservation. When not in the nursery, she has been found leading tropical nature tours in Costa Rica and natural history tours in Southeast Alaska.

Andrew Merritt's appreciation for native plants began in his youth in New England when he encountered his first pink lady's slipper (*Cypripedium acaule*). His passion for plants grew as he worked as a gardener and landscaper. After years of traveling, Andrew settled on the Oregon side of the Columbia River Gorge where his love for native plants bloomed. Andrew met his wife, Kristin, and together they created Humble Roots Nursery, a native plant nursery specializing in the plants of the Columbia River Gorge and the Pacific Northwest. Through Humble Roots, Andrew and Kristin have worked on many restoration and pollinator enhancement projects including rare plant monitoring and propagation. Andrew works with homeowners, landscapers, farmers, orchardists, organizations, and agencies developing native plant gardens and habitats. He enjoys his days collecting seeds and propagating plants from the diverse array of the region's flora.